Lecture Notes in Artificial Inte

Subseries of Lecture Notes in Computer S
Edited by J. G. Carbonell and J. Siekmann

Lecture Notes in Computer Science

Edited by G. Goos, J. Hartmanis, and J. van Leeuwen

Springer
Berlin
Heidelberg
New York
Barcelona
Hong Kong
London
Milan
Paris
Tokyo

David J. Hand   Niall M. Adams
Richard J. Bolton (Eds.)

Pattern Detection and Discovery

ESF Exploratory Workshop
London, UK, September 16-19, 2002
Proceedings

 Springer

Series Editors

Jaime G. Carbonell, Carnegie Mellon University, Pittsburgh, PA, USA
Jörg Siekmann, University of Saarland, Saarbrücken, Germany

Volume Editors

David J. Hand
Niall M. Adams
Richard J. Bolton
Imperial College of Science, Technology and Medicine
Department of Mathematics
Huxley Building, 180 Queen's Gate
London, SW7 2BZ, UK
E-mail: {d.j.hand, n.adams, r.bolton}@ic.ac.uk

Cataloging-in-Publication Data applied for

Die Deutsche Bibliothek - CIP-Einheitsaufnahme

Pattern detection and discovery : ESF exploratory workshop, London, UK,
September 16 - 19, 2002 / David J. Hand ... (ed.). - Berlin ; Heidelberg ;
New York ; Barcelona ; Hong Kong ; London ; Milan ; Paris ; Tokyo :
Springer, 2002
   (Lecture notes in computer science ; Vol. 2447 : Lecture notes in
   artificial intelligence)
   ISBN 3-540-44148-4

CR Subject Classification (1998): I.2, H.2.8, F.2.2, E.5, G.3, H.3

ISSN 0302-9743
ISBN 3-540-44148-4 Springer-Verlag Berlin Heidelberg New York

Springer-Verlag Berlin Heidelberg New York,
a member of BertelsmannSpringer Science+Business Media GmbH

http://www.springer.de

© Springer-Verlag Berlin Heidelberg 2002
Printed in Germany

Typesetting: Camera-ready by author, data conversion by PTP-Berlin, Stefan Sossna e.K.
Printed on acid-free paper   SPIN: 10871071   06/3142   5 4 3 2 1 0

Preface

The collation of large electronic databases of scientific and commercial information has led to a dramatic growth of interest in methods for discovering structures in such databases. These methods often go under the general name of data mining. One important subdiscipline within data mining is concerned with the identification and detection of anomalous, interesting, unusual, or valuable records or groups of records, which we call *patterns*. Familiar examples are the detection of fraud in credit-card transactions, of particular coincident purchases in supermarket transactions, of important nucleotide sequences in gene sequence analysis, and of characteristic traces in EEG records. Tools for the detection of such patterns have been developed within the data mining community, but also within other research communities, typically without an awareness that the basic problem was common to many disciplines. This is not unreasonable: each of these disciplines has a large literature of its own, and a literature which is growing rapidly. Keeping up with any one of these is difficult enough, let alone keeping up with others as well, which may in any case be couched in an unfamiliar technical language. But, of course, this means that opportunities are being lost, discoveries relating to the common problem made in one area are not transferred to the other area, and breakthroughs and problem solutions are being rediscovered, or not discovered for a long time, meaning that effort is being wasted and opportunities may be lost.

The aims of this workshop were to draw together people from the variety of disciplines concerned with this common interest, and to attempt to characterize more soundly the fundamental nature of their common interests. That is, (i) we sought to break down barriers, so that advances in one area could be communicated to others, and so that the common nature of the efforts could be recognized and taken advantage of, and (ii) we aimed to distil the essence of the common problem of pattern detection, so that the discipline could advance on a sound footing.

The various literatures concerned with pattern detection have arisen from practical needs. Perhaps inevitably because of this, there has been a heavy emphasis on the development of algorithms and methodology, with very little attention being paid to the development of a sound theory. But a sound theoretical basis is important, if one is to characterize those areas which need new research, if one is to identify strategies for problems thought to be impossible, and if one is to be able to transfer methodology between entirely different application areas. Perhaps above all, a sound theory is important if one is to tackle the problems which bedevil pattern detection, such as the occurrence of spurious patterns by chance alone, patterns arising from data distortion, and the issue of patterns which, though real, are of no practical significance or which are already well-known. Without addressing issues such as these, pattern detection (and data mining more broadly) is likely to suffer a backlash as users find that the inflated promises are not being fulfilled.

The workshop was made possible by generous funding from several sources:

- The European Science Foundation provided the bulk of the funding. Without its support, this workshop would not have been possible.
- The Royal Statistical Society saw the value of what we were trying to do, and its importance to statisticians, and to statistics as a discipline.
- Barclaycard recognized the relevance of our aims for organizations such as theirs.
- Fair, Isaac also saw the potential importance of what we were trying to do.

To all these organizations we are very grateful.

July 2002

David J. Hand
Richard J. Bolton
Niall M. Adams

Participants at the Workshop

Niall Adams
Department of Mathematics, Imperial College, UK
n.adams@ic.ac.uk

Helena Ahonen-Myka
Department of Computer Science, University of Helsinki, Finland
helena.ahonen-myka@cs.helsinki.fi

Richard J. Bolton
Department of Mathematics, Imperial College, UK
r.bolton@ic.ac.uk

Jean-Francois Boulicaut
INSA-Lyon Scientific and Technical University, France
jfboulic@lisi.insa-lyon.fr

Paul Cohen
Experimental Knowledge Systems Laboratory, Department of Computer Science,
University of Massachusetts, USA
cohen@cs.umass.edu

Ursula Gather
Fachbereich Statistik, University of Dortmund, Germany
gather@statistik.uni-dortmund.de

Bart Goethals
School for Information Technology, University of Limburg (tUL), Belgium
bart.goethals@luc.ac.be

David J. Hand
Department of Mathematics, Imperial College, UK
d.j.hand@ic.ac.uk

Frank Höppner
Department of Computer Science, University of Applied Sciences BS/WF,
Germany
hoeppner@et-inf.fho-emden.de

Xiao Hui Lui
Department of Information Systems and Computing, Brunel University, UK
xiaohui.liu@brunel.ac.uk

Marzena Kryszkiewicz
Institute of Computer Science, Warsaw University of Technology, Poland
mkr@ii.pw.edu.pl

Heikki Mannila
Department of Computer Science, University of Helsinki, Finland
Heikki.Mannila@hut.fi

Dunja Mladenić
Department for Intelligent Systems, J. Stefan Institute, Slovenia
Dunja.Mladenic@ijs.si

Katharina Morik
Fachbereich Informatik, University of Dortmund, Germany
morik@ls8.cs.uni-dortmund.de

Pierre-Yves Rolland
Atelier de Modélisation et de Prévision, Université d'Aix-Marseille III, France
p_y_rolland@yahoo.com

Daniel Sánchez
Department of Computer Science and Artificial Intelligence (DECSAI),
University of Granada, Spain
daniel@decsai.ugr.es

Arno Siebes
Centrum voor Wiskunde en Informatica, Amsterdam, The Netherlands
Arno.Siebes@cwi.nl

Myra Spiliopoulou
Department of E-Business, Leipzig Graduate School of Management (HHL),
Germany
myra@ebusiness.hhl.de

Antony Unwin
Institut für Mathematik, Universität Augsburg, Germany
Antony.Unwin@Math.Uni-Augsburg.de

Marek Wojciechowski
Politechnika Poznanska, Instytut Informatyki, Poland
Marek.Wojciechowski@cs.put.poznan.pl

Chris Workman
Center for Biological Sequence Analysis, Institute of Biotechnology,
The Technical University of Denmark, Denmark
workman@cbs.dtu.dk

Organization

The ESF Exploratory Workshop on Pattern Detection and Discovery was organized by the Department of Mathematics, Imperial College of Science, Technology and Medicine.

Executive Committee

Chair: David J. Hand, Imperial College of Science, Technology and Medicine

Organization: Richard J. Bolton, Imperial College of Science, Technology and Medicine

Publication: Niall M. Adams, Imperial College of Science, Technology and Medicine

Referees

Niall Adams
Richard Bolton
David Hand
Yannis Manolopoulos
Nicolas Pasquier
Graeme Richards

Sponsoring Institutions

European Science Foundation
The Royal Statistical Society
Barclaycard
Fair, Isaac

Table of Contents

Pattern Detection and Discovery

Text and Web Mining

Applications

Pattern Detection and Discovery

David J. Hand

Department of Mathematics, Imperial College
180 Queen's Gate, London, UK. SW7 2BZ
d.j.hand@ic.ac.uk
http://stats.ma.ic.ac.uk/

Abstract. Data mining comprises two subdisciplines. One of these is based on statistical modelling, though the large data sets associated with data mining lead to new problems for traditional modelling methodology. The other, which we term *pattern detection*, is a new science. Pattern detection is concerned with defining and detecting local anomalies within large data sets, and tools and methods have been developed in parallel by several applications communities, typically with no awareness of developments elsewhere. Most of the work to date has focussed on the development of practical methodology, with little attention being paid to the development of an underlying theoretical base to parallel the theoretical base developed over the last century to underpin modelling approaches. We suggest that the time is now right for the development of a theoretical base, so that important common aspects of the work can be identified, so that key directions for future research can be characterised, and so that the various different application domains can benefit from the work in other areas. We attempt describe a unified approach to the subject, and also attempt to provide theoretical base on which future developments can stand.

1 Introduction

Researchers in bioinformatics try to find nucleotide patterns which have important influences on the organism. Researchers in speech recognition try to detect when certain patterns of sound signals appear. Those involved with crime detection try to spot when particular patterns of credit card transactions occur. People who work for hedge funds try to spot anomalous behaviour in the financial markets. All of these people, and a host of others, are concerned not with modelling data in the way developed by statisticians, but with detecting small configurations of data values which are anomalous, interesting, or of special significance in some other way. That is, all of these people have a common aim, even though this aim is often concealed by the very different languages and interests of the areas in which they are working. In this paper, I shall make an attempt to extract the common nature of this activity and to define the concepts and terms involved.

I give a more formal definition of what I mean by a pattern in Section 2, but to get us going, let us informally define a pattern as a configuration of data

D.J. Hand et al. (Eds.): Pattern Detection and Discovery, LNAI 2447, pp. 1–12, 2002.

values which is of special interest. We note immediately that this is different from other usages of 'pattern' - the word is used in different ways in different situations. In particular, we are not referring to repetitive regularities - as in wallpaper pattern (e.g., [1]).

In 'pattern matching', the characterisation of what is of interest will come from outside the data. Thus we may be seeking instances of a particular word being spoken, or occasions when airline tickets and train tickets are purchased together, or patients in whom three particular symptoms occur simultaneously. In each of these cases we have been told beforehand what configuration, what pattern, we are seeking. Given the characterisation or definition of configurations of interest, in pattern matching our aim is to locate occurrences of that pattern in the database - for various possible reasons.

In 'supervised pattern detection', the characterisation of what is of interest comes from a particular response or classification variable in the data. Our aim is then to locate patterns, configurations, in the data that maximise (minimise) a response variable or minimise some misclassification error. For example, we might be interested in finding patterns in a database of geodemographic data that maximise a person's spending on a particular product. An indicator variable representing fraudulent/non-fraudulent purchases can be used to detect patterns of credit card spending that indicate fraud. Algorithms that perform supervised pattern detection include Klösgen's Explora subgroup mining system [2] and Friedman and Fisher's SuperGEM patient rule induction method [3]. We have stolen the word 'supervised' from 'supervised pattern recognition', where we are told to which of a possible set of classes each object (in a 'design' set) belongs. 'Supervised' because this information comes from outside - from a 'supervisor'.

In contrast, in 'unsupervised pattern detection', we not only have to locate occurrences of the patterns in the database, but we also have to decide, solely using information in the database, whether such an occurrence is a pattern. If the definition of a pattern has to be based on internal database information, it can only refer to relationships between values in individual data vectors, relationships between different data vectors, or the two together. In this paper, I am especially concerned with unsupervised pattern detection. The problem of pattern matching seems to me to reduce to questions of defining appropriate distance measures and developing efficient search and matching algorithms. It does not seem to me that such areas involve deep conceptual issues about what a pattern 'is'. On the other hand, there is still a substantial advantage to be gained by technology transfer, adopting algorithms developed in one area of supervised pattern detection for use in other areas. Although the data arising in different areas will have different structural features (for example, in many areas the data are sequential, in others data points are embedded in two or three dimensional manifolds, in others there are trellis relationships between different points) it is likely that advances in algorithm development in one area will benefit other areas.

Pattern detection as a defined area of research in its own right seems relatively new. This might be regarded as odd, because so much of our knowledge is couched in terms of patterns. For example, we know that 'people who smoke are more likely to suffer from lung cancer', that 'simultaneous use of a credit card in two different geographical locations suggests fraud', and that a star which appears to move against the background of around 20 million stars suggests gravitational microlensing and the presence of a black hole. These examples might seem to suggest that pattern detection is nothing but rule detection. Expert systems, which were the focus of so much research and enthusiasm in the 1980s, were primarily based on rule-system architectures. In fact, most of those systems were based on knowledge – expertise – extracted from human experts, rather than on rules distilled from databases. The problem of knowledge elicitation has proven to be one of the difficulties in implementing such systems. In any case, pattern detection is more general than this. The antecedent-consequent structure is a defining characteristic of a rule, but not of a pattern. Thus, our suspicions are aroused if several people fall ill with a stomach bug: we begin to search for the common cause. Similarly, will try to identify small groups of people who use their credit card in a characteristic way: perhaps later we could use this information for commercial gain. Neither of these patterns are rules.

If it is odd that appreciation of pattern detection as a discipline in its own right is so new, perhaps its novelty can be explained by the technology required to support the exercise. Pattern detection requires the ability to search through a large, often vast, set of possible patterns, looking for relationships between the variables within data points and also relationships between distinct data points. For databases other than the very smallest, this sort of power requires fast computers - and, in any case, anomalous data configurations are unlikely to be found in small databases. These properties distinguish pattern detection from model building, the other core exercise within data mining (see, for example, [4], [5]). Model building is based on ideas developed by statisticians, primarily over the twentieth century. Although large data sets pose new modelling challenges, and although computers have led to the development of entirely new kinds of modelling tools, modelling ideas were originally developed on small data sets, data sets where the ideas could be applied by hand. Statistics began before the computer age. Because of this, statistics was originally a mathematical discipline (though I have argued that it should nowadays be more properly regarded as a computational discipline) and a consequence of this has been an emphasis on sound formalism and carefully thought out inferential ideas. In contrast, the pattern detection literature, with clear origins in the computational disciplines which provided the huge databases and which could provide the fast search algorithms to examine the databases, has focused on the computational aspects, and on algorithms in particular. A glance at the *Proceedings* of any of the major data mining conferences will demonstrate this. There has been a dearth of solid theoretical development, analogous to statistical theory for modelling, to underpin work on pattern detection. (Statistical work on outlier detection, a kind of unsupervised pattern detection, is a rare and noticeable exception.) There has been plenty of work on the *methodology* of pattern detection, but very limited work on the theory.

Does this matter? I would argue that it is crucial. Pattern detection, like statistical modelling, is about *inference*. We are aiming to detect those data configurations which are due to real phenomena underlying the data generating process, and not those configurations which have arisen by chance. Without an underlying theory, how can one do this? Without an underlying theory, how does one even know precisely what one means by a 'pattern' - and how the concepts and tools generalise and transfer between domains. Beyond this, a strong theoretical foundation permits one to identify which tools are useful for what purposes, and to assess and compare the performance of such tools, as well as helping to identify which areas merit further research. Without this, one is making a mistake precisely similar to that still occasionally made in elementary statistics teaching: the *cookbook fallacy*. This regards statistical methods as distinct tools, one to be chosen to match a given data set and problem. It fails to appreciate the interconnected structure of statistical ideas and methods; the pattern of generalisation and specialisation which binds the tools together into a single whole. It is this essential interconnectedness which makes statistics so powerful. It enables new tools to be developed to meet the problems of new data sets, permitting one to be confident that sound inferential conclusions can be drawn. Without an appreciation of this framework, a new problem leaves one floundering in the dark.

In summary, the rapid and dramatic growth of pattern detection ideas, in a wide variety of application domains, has been driven primarily by methodological requirements. Development of a theoretical base has lagged behind, and largely still remains to be done. It is the aim of this paper to suggest a way in which such a base might be established.

2   What Is a Pattern?

Here is an initial stab at what we might mean by 'pattern' in unsupervised pattern detection: *a pattern is a data vector serving to describe an anomalously high local density of data points.*

The word 'anomalous' implies that there is something to which the local density of data points can be compared. That is, there is some kind of 'background', 'baseline', or 'expected' value for the density. This will be a statistical model of some kind, and will be more or less elaborate according to whether a lot or a little is known about the data source. For example, if very little is known, one might simply postulate some kind of uniform background model. In some form or other, sometimes implicitly rather than explicitly, this is the background model which has been adopted by [6], [7], [8], [9] and [10]. A slightly more sophisticated background model is one which assumes independence between variables - used, for example, in [11], [12] and [13]. This is the model used in much nucleotide sequence analysis, where the baseline probability that a particular nucleotide will occur at any point is simply the overall proportion of occurrences of this particular nucleotide in the data. Independence is often relaxed by introducing a Markov assumption - especially in domains concerned with sequential data such as nucleotide sequence analysis and speech recognition. See, for example, [14], [15], [16] and [17].

Since a pattern represents departure from a background model, we can represent our description of the data as

$$data = background\_model + pattern + random\_component$$

This is an extension of the classical statistical modelling perspective, which simply includes the *background_model* and *random_component* parts. The important thing about this representation is that it makes the role of the background model explicit. Many of the application domains in which pattern detection algorithms have been developed made no explicit mention or acknowledgement of the necessity for a background model - and perhaps this explains the lack of a unifying theoretical development.

Ideally, the background model will be based on solid ideas about the data and the data generating process. However, other desirable attributes of the background model ([13]) are that it should be easily interpretable (so that deviations from it are easily interpretable), and that it should not be sensitive to the particular set of data available.

The patterns found in unsupervised pattern detection are essentially syntactic - they are identified purely on the basis of data configurations relative to the background, and not on what these configurations might mean. (However, the background model itself will have semantic content, so the semantic degree of a pattern might be defined relative to that - see [18]) Of course, what is of interest - and this is really the point of unsupervised pattern detection - is the *semantic* implication of a pattern. Having found a structure in the data, we want to know what it means. At the least we will want to know, that the pattern has not arisen because of distortion in the data (missing values being regarded as genuine data values, for example) and is unlikely to be attributable to random variation.

The definition referred to a *local* density of data points. This is meant to characterise a pattern as being a 'small' phenomenon in some sense. A cluster analysis may decompose a data set into regions of high point density, but together these will decompose the data set into mutually exclusive subsets, some of which may be large. As such, they would not be patterns. Of course, it may sometimes be possible to decompose a data set into many small local clusters, and then each might be legitimately regarded as a pattern. Such an example would represent the interface or overlap of modelling and pattern detection ideas. Hand and Bolton [18] give the example of natural language processing, in which each spoken word might be regarded as a pattern to be detected, but in which an entire utterance is split into such patterns. In order to define what is meant by 'local' it is necessary to adopt a suitable distance measure. This choice will depend on the data and the application domain: in the cases of categorical variables, it might even require exact matches. It is also necessary to pick a threshold with which the measured distance is compared. If 'local' is taken to include the entire data set, then a single point will be identified as a pattern: the global maximum. At the other extreme, if 'local' is taken to be a very small neighbourhood, then many slight irregularities in the estimated point density will be detected as patterns.

Some sort of compromise is needed. One way to find a suitable compromise would be to gradually expand the distance threshold defining local, so that the set of points identified as possible patterns gradually increases, stopping when the detected number of patterns seemed reasonable (a decision which depends on resources as well as on the phenomenon under investigation).

I have suggested summarising a locally dense region by a single point, the pattern. One might, for example, pick the point corresponding to the greatest local density, as estimated by some nonparametric density estimation procedure such as a kernel method. Greatest, here, is important. If we simply compared the density with a threshold value, then (a) large entire regions might qualify, and (b) neighbouring points to the 'greatest' would certainly qualify. We are really concerned with finding local maxima, where 'local' is restricted to small regions. Point (b) here is sometimes referred to as the *foothill effect* ([19], [18]). The practice of choosing a *single* 'best' (in some sense) representative point from a local region is related to the notion of *minimal cover rules* ([20]). Another related but rather different phenomenon is that arising from the fact that patterns will have subpatterns. Indeed, one of the basic pattern detection algorithms, the *a priori* rule, is based on the fact that a given set of data values cannot exceed a given threshold frequency if all of its subsets do not also exceed that threshold.

3   Detecting Patterns

From the preceding section, we can see the basic principle which must lie behind the detection of patterns: we will estimate the local density of data points and compare this density with the density we would expect on the basis of the background model. If the difference exhibits a local peak, then we flag this peak as a possible pattern. All of the many algorithms we have seen use this basic strategy, even if they do not make explicit the background model against which the smoothed data are being compared. They use various measures, depending on the context, to determine if the local anomaly is large enough to be of interest. For example, in association analysis measures of confidence and support (equivalent to estimates of conditional and marginal probabilities) are used. In genome sequence analysis, estimates of the probability of observing a given sequence under the background model are calculated. These measures are then compared with a threshold to choose those thought to merit more detailed examination (subject to the restriction that not too many must be proposed - see [21]. DuMouchel [13] proposes a Bayesian approach to ranking cells of a contingency table in terms of their deviation from an independence model, based on assuming that the count in the cells follows Poisson distributions with parameters distributed according to a prior which is a mixture of two gamma distributions.

Patterns are defined as local deviations from a background model. However, the data have arisen from a stochastic process, so we should expect deviations from the model to arise purely by chance. Some, indeed, should be expected to be large. Moreover, since there are many possibilities for such deviations to occur, especially when the data sets are large, we should expect to find some.

How, then, can we decide whether or not such deviations represent aspects of the underlying mechanism rather than mere chance fluctuations?

Standard statistical tools have been developed for tackling analogous problems in different contexts. For example, the Bonferroni adjustment and others (e.g. [22], [23]) control the *familywise error rate*, the probability of incorrectly identifying a pattern. However, they achieve this at the cost of very low power for detecting individual false null hypotheses - in our terms they will not be very effective at detecting true anomalies in the underlying data generating mechanism. In any case, perhaps this is not quite the right strategy: rather than controlling the proportion of 'false patterns' which are flagged as worth investigation, we should control the proportion of those patterns flagged as worth investigation which are 'false' - which do not represent true anomalies in the underlying data generating mechanism. This, the flagging of spurious patterns, is, of course, what generates unnecessary and unrewarding work (and is what may lead to a backlash against the entire exercise when only a tiny minority of the patterns one has flagged turn out to be of any value). Unfortunately, this cannot be determined because we cannot tell, in any real situation, which of the flagged patterns actually do correspond to a true anomaly in the underlying data generating mechanism. What we can do, however, following the ideas of Benjamini and Hochberg [24] is control the expected proportion of flagged patterns which do not correspond to true anomalies in the underlying data generating mechanism.

We will use the notation in Table 1 to explore this. To simplify the terminology, we will define a pattern as 'substantive' if it represents a real structure in the distributions from which the data were generated. Otherwise it is 'nonsubstantive'. For simplicity we will suppose $d > 0$ though the result can be extended to include the $d = 0$ case. Let

$$I(b > 0) = \begin{cases} 0 \; if \; b = 0 \\ 1 \; if \; b > 0 \end{cases}$$

Then

$$\frac{b}{b+d} = I(b > 0) \quad if \; b = 0$$

and

$$\frac{b}{b+d} < I(b > 0) \quad if \; b > 0$$

so that, in general,

$$\frac{b}{b+d} \leq I(b > 0)$$

Taking expectations, we have

$$E\left(\frac{b}{b+d}\right) \leq E(I(b > 0)) = P(b > 0)$$

That is, the expected proportion of the flagged patterns which are not substantive is bounded above by the familywise error rate. The implication is that controlling the expected proportion of the flagged patterns which are not substantive might lead to a less conservative result than the traditional multiple testing procedures. Benjamini and Hochberg [24] give a simple algorithm for controlling the expected proportion of the flagged patterns which are not substantive at any chosen level.

This tool was originally developed for applications other than pattern detection, and it remains to be seen how effective it is in this new application. Results from Bolton *et al* [25] suggest that it may not always be very effective. The point is that Benjamini and Hochberg [24] originally developed the idea for testing multiple hypotheses in situations in which one might reasonably expect a substantial number of the hypotheses to be 'substantive' (using our term). This means that, in their situation, $b/(b+d)$ may be markedly less than unity. In pattern detection, however, one might expect (or perhaps hope) that there will be few substantive patterns and many non-substantive patterns. That would give $b/(b+d) \approx 1$, so that $E(b/(b+d))$ would not be a great deal less than $P(b > 0)$.

Table 1. Frequencies of substantive and flagged patterns.

	Not flagged	Flagged
Non-substantive	a	b
Substantive	c	d

Controlling familywise error rate and controlling the expected proportion of the flagged patterns which are not substantive are two strategies, based on standard frequentist arguments, which might be adopted in attempting to ensure that one does not identify too many patterns as worth investigating in detail. However, perhaps a more appropriate approach is based on likelihood argument. Significance test approaches control the probability of drawing incorrect conclusions (either that a substantive pattern exists when one does not, or vice versa). However, as far as pattern detection is concerned, we might assert that we are not really interested in such probabilities, but rather in the *strength of evidence* that the data provide for the existence of a substantive pattern. Relative strength of evidence is given not by type 1 or 2 error probabilities, but simply by likelihood ratios. Similarly, Bayesian strategies combine the likelihood with the priors, thus combining the evidence in the data with prior beliefs. This is appropriate if the aim is to decide how to adjust one's beliefs, but not if the aim is simple to indicate the extent of evidence which the data provide for departures from a background model. Procedures such as those described above are necessary to control error probabilities, but no such procedures are necessary when one is examining evidence. In particular, the evidence in favour of the existence of a substantive pattern is not influenced by whether there is other evidence favouring some other pattern. With the likelihood approach, for each of the separate patterns being examined, we can determine a likelihood interval. If the observed pattern lies outside this interval, then there is evidence (at a level we determine) favouring the existence of a substantive pattern.

To illustrate, suppose that the background model is the independence model. For illustration, suppose that the data are categorical, and assume that, under the background model each cell contains a number of data points which follows a Poisson distribution with parameter given by the background model for that cell (see, e.g., [13]). Then the evidence in a set of data about hypothesis A compared to hypothesis B is given by the ratio of the two likelihoods under these hypotheses: L(A)/L(B). For a cell with observed count x, the likelihood is $L(\lambda) = \frac{e^{-\lambda}\lambda^x}{x!}$. The maximum of this is $\frac{e^{-x}x^x}{x!}$. For all values of λ for which $\frac{e^{-x}x^x}{x!} \Big/ \frac{e^{-\lambda}\lambda^x}{x!} > k$ there is at least one other value that is better supported by a factor of at least k. (A value of $k = 20$ would be a reasonable choice.) We therefore flag, as worth investigating, all those cells which have expected counts E satisfying $e^{-x}x^x > ke^{-E}E^x$. Note that since we are only interested in situations where the count is larger than the expected count we should restrict ourselves to expected counts E which also satisfy $E < x$. If the expected count E for a cell, calculated from the background model, satisfies these conditions, then there are alternative probability models, based on local peaks, which have a substantially greater likelihood than the background model.

4   Data Quality

The aim of pattern detection is typically to detect features of the underlying data generating process. It is not to detect local aggregations of data points which have arisen by chance (Section 3), and it is not (normally) to detect local aggregations of data points which have arisen due to distortion in the data collection or recording processes. But in some applications many, possibly most, of the detected patterns will be attributable to data collection problems. (Perhaps we should remark here that detecting such problems can, of course, be invaluable, if it leads to the later rectification of those problems.) The discovery by Dave Yearling (one of my PhD students) that the apparent gales in windspeed records were due to automatic instrument resetting does not tell us anything of value about meteorology. The discovery by Berry and Linoff [26] that 20% of transactions involved files transmitted before they arrived, due to failure to reset the computer's clock, tells us little about human behaviour. The list of errors in coding of birth weights noted by Brunskill [27], with 14oz being recorded as 14lb, birth weights of one pound (1lb) being read as 11lb, and misplaced decimal points (for example, 510 gms recorded as 5100 gms), casts doubt on the value of the apparent excess of overweight infants at low gestational ages.

Data quality is a *key* issue in pattern detection. It is very much a practical rather than a theoretical issue, but it is vital. We are reminded of Twyman's Law: *any figure that looks interesting or different is usually wrong.*

5   Algorithms

The bulk of the work on pattern detection has focused on the development of algorithms - on methodology - rather than theory. This is understandable: algorithms are the sharp end of pattern detection. Indeed, in order to be able to develop a theory, in order to understand what the problems and issues are which require theoretical solution, it is necessary to have gained some practical experience. However, our aim at this workshop is to move the discipline up a level. Indeed, continuing the metaphor, one might say that our aim is to move the discipline up a level and insert an underpinning theoretical foundation beneath it. A vast range of algorithms have been developed, many relying on common principles, and others taking advantage of special properties of the data sets to which they will be applied. They have evolved in response to the continuing demands of increasing size and speed as data collection and storage capacities continue to grow.

The basic starting point is, of course, exhaustive search, but this soon becomes impossible as the data set and number of combinations of data points grows. Another key advance, applicable in many areas (such as association analysis) was the *apriori algorithm*. In fact, this algorithm provides a nice illustration of the need for a deeper theoretical basis. It is based on the principle that a pattern A cannot have a frequency above some threshold if some subpattern of A has a frequency below this threshold, and has served as the basis (and comparator) for a large amount of algorithm development. However, it ignores the fact that a probabilistic or likelihood threshold should depend on the background distribution, and that this will be different for A and subpatterns of A. It is entirely possible that A may pass its relevant threshold while all subpatterns of A fail to pass theirs.

Other aspects of data structure are made use of in methods such as branch and bound, sequential forward and backward stepwise algorithms, more general strategies for concatenating subcomponents, and so on. The complexity of the problem, and the size of the search spaces, increases astronomically when wildcards or variable length patterns are permitted. Sampling does have a role to play, and classical statistical notions become important in calculating the probability that patterns will fail to be detected if one is merely using a sample of the data. Perhaps it goes without saying that stochastic search methods are also applicable in this context.

6   Conclusion

Pattern detection has become an important discipline in its own right, having grown up as a subdiscipline of data mining, but having also appeared in several unconnected application domains, including text analysis, data compression, genomics, speech recognition, and technical analysis. The thrust of the work to date has been towards developing algorithms for practical application; that is, almost all of the work to date has been concerned with *methodology*. This has left something of a hole at the core of the discipline - indeed, even a lack of awareness that there was a common aim, and that there was a single discipline concerned. This historical development is perhaps natural, since inevitably practical applications have provided the incentive for work in the area. However, the time is now right to step back and take a more considered view of work in the area, and to underpin it with a stronger theoretical base: to develop a *theory* of pattern detection. Such a theory allows one to critically assess and evaluate work which is being done, and to decide what problems are important. The development of a unified perspective also allows one to transfer ideas and tools from one domain to another, with mutual benefit. In general, in science, theory and methodology perform a leapfrogging act, progress in one permitting progress in the other, which then feeds back to the first, and so on. This is perhaps most clearly illustrated in the physical and modern biological sciences, by the leapfrog act between theory and instrumentation. Now is the time for the theory of pattern detection to catch up with and feed back into the practice.

We have presented a possible approach to a theory of pattern detection in this paper. It may not be the only approach, and many questions clearly remain open. We hope, however, that it will stimulate further work.

Acknowledgements. The work described in this paper was supported by EP-SRC ROPA grant GR/N08704.

References

1. Grenander U.: General Pattern Theory: a Mathematical Study of Regular Structures. Clarendon Press, Oxford (1993)
2. Klösgen, W.: Subgroup patterns. In: Klösgen, W., Zytkow, J.M. (eds.): Handbook of data mining and knowledge discovery. Oxford University Press, New York (1999)
3. Friedman, J.H., Fisher, N.I.: Bump hunting in high-dimensional data. Statistics and Computing **9(2)** (1999) 1–20
4. Hand D.J., Blunt G., Kelly M.G., Adams N.M.: Data mining for fun and profit. Statistical Science **15** (2000) 111–131
5. Hand D.J., Mannila H., Smyth P.: Principles of Data Mining. MIT Press (2001)
6. Chau T., Wong A.K.C.: Pattern discovery by residual analysis and recursive partitioning. IEEE Transactions on Knowledge and Data Engineering **11** (1999) 833–852
7. Adams N.M., Hand D.J., Till, R.J.: Mining for classes and patterns in behavioural data. Journal of the Operational Research Society **52** (2001) 1017–1024
8. Bolton R.J., Hand D.J.: Significance tests for patterns in continuous data. In: Proceedings of the IEEE International Conference on Data Mining, San Jose, CA. Springer-Verlag (2001)
9. Edwards R.D., Magee F.: Technical Analysis of Stock Trends. 7th edn. AMACOM, New York (1997)
10. Jobman D.R.: The Handbook of Technical Analysis. Probus Publishing Co. (1995)
11. Zembowicz R., Zytkow J.: From contingency tables to various forms of knowledge in databases. In: Fayyad, U.M., Piatetsky-Shapiro, G., Smyth, P., Uthurusamy, R. (eds.): Advances in Knowledge Discovery and Data Mining, Menlo Park, California, AAAI Press (1996) 329–349

12. Liu B., Hsu W., Ma Y.: Pruning and summarizing the discovered associations. In: Proceedings of the Fifth ACM SIGKDD International Conference on Knowledge Discovery and Data Mining, San Diego, California, ACM Press (1999) 125–134

13. DuMouchel, W.: Bayesian data mining in large frequency tables, with an application to the FDA Spontaneous Reporting System. The American Statistician **53** (1999) 177–202

14. Jelinek F.: Statistical Methods for Speech Recognition. MIT Press, Cambridge, Massachusetts (1997)

15. Sinha S., Tompa M.: A statistical method for finding transcription factor binding sites. In: Proceedings of the Eighth International Conference on Intelligent Systems for Molecular Biology, La Jolla, CA, AAAI Press (2000) 344–354

16. Chudova, D., Smyth, P.: Unsupervised identification of sequential patterns under a Markov assumption. In: Proceedings of the KDD 2001 Workshop on Temporal Data Mining, San Francisco, CA (2001)

17. Durbin R., Eddy S., Krogh A., Mitchison G.: Biological Sequence Analysis. Cambridge University Press: Cambridge (1998)

18. Hand D.J., Bolton R.J.: Pattern detection in data mining. Technical Report, Department of Mathematics, Imperial College, London (2002)

19. Dong G., Li J.: Interestingness of discovered association rules in terms of neighbourhood-based unexpectedness. In: Proceedings of the Pacific Asia Conference on Knowledge Discovery in Databases (PAKDD), Lecture Notes in Computer Science, Vol. 1394., Springer-Verlag, Berlin Heidelberg New York (1998) 72–86

20. Toivonen H., Klemettinen M., Ronkainen P., Hätönen, Mannila H.: Pruning and grouping discovered association rules. In: Mlnet Workshop on Statistics, Machine Learning, and Discovery in Databases, Crete, Greece, MLnet (1995) 47–52

21. Brin S., Motwani R., Ullma J.D., Tsur S.: Dynamic itemset counting and implication rules for market basket data. In: Proceedings of the ACM SIGMOD International Conference on Management of Data, Tucson, Arizona, ACM Press (1997) 255–264

22. Miller R.G.: Simultaneous Statistical Inference. 2nd ed. Springer-Verlag, New York (1981)

23. Pigeot I.: Basic concepts of multiple tests – a survey. Statistical Papers **41** (2000) 3–36

24. Benjamini Y., Hochberg Y.: Controlling the false discovery rate. Journal of the Royal Statistical Society, Series B **57** (1995) 289–300

25. Bolton R.J., Hand D.J., Adams, N.: Determining hit rate in pattern search. In: These Proceedings (2002)

26. Berry M.J.A., Linoff G.: Mastering data mining. The art and science of customer relationship management. Wiley, New York (2000)

27. Brunskill A.J.: Some sources of error in the coding of birth weight. American Journal of Public Health **80** (1990) 72–3

Detecting Interesting Instances

Katharina Morik

Univ. Dortmund, Computer Science Department, LS VIII
morik@ls8.informatik.uni-dortmund.de,
http://www-ai.cs.uni-dortmund.de

Abstract. Most valid rules that are learned from very large and high dimensional data sets are not interesting, but are already known to the users. The dominant model of the overall data set may well suppress the interesting local patterns. The search for interesting local patterns can be implemented by a two step learning approach which first acquires the global models before it focuses on the rest in order to detect local patterns. In this paper, three sets of interesting instances are distinguished. For these sets, the hypothesis space is enlarged in order to characterize local patterns in a second learning step.

1 Introduction

Most valid rules that are learned from very large and high dimensional data sets are not interesting, but are already known to the users. For instance, we have learned from a very large data set on cars and their warranty cases that the production date of a car is preceding its sales date, which is what we expected [14]. However, there were some exceptions to the rule and these are interesting. Which customers order cars before they are offered? Are the exceptions typing errors? The decision can only be made by domain experts. Either we use the outliers for data cleaning or we find interesting instances by learning a general rule and outputing its exceptions. This idea has been put forward by Vladimir Vapnik [7]. In some data bases, we are looking for fraudulent use of cellular phones as has been reported by Fawcett and Provost [4]. There, finding the instances that are exceptions to generally reliable rules corresponds to finding the criminal indiviuals. In other applications we learn general models for customer behavior, but are interested in exceptionally lucrative customers. Again, these customers are found by collecting the exceptions to accepted rules. Hence, one reason for a set of instances being interesting is that they are exceptions to a general model.

Many learning algorithms aim at covering all positive examples. However, there might be some instances that cannot be covered by any rule. Why do they not fit into any general model? What makes them so exceptional? Maybe these uncovered instances are the most interesting ones and lead us to a local pattern.

A third phenomenon can again be illustrated by the database of cars and their parts. We encountered valid rules not being learned. For instance, it could not be learned that each car has at least one axle. Since we know that this rule should be

D.J. Hand et al. (Eds.): Pattern Detection and Discovery, LNAI 2447, pp. 13–23, 2002.
© Springer-Verlag Berlin Heidelberg 2002

verified by the data, we inspected the evidence against the rule and found many missing attribute values in the database. We could clean the data base on the basis of the refutation of valid rules. In a medical application, we looked for effects of drugs. We knew from the domain expert that a particular drug should decrease blood pressure. However, the rule was not learned that after the intake of the drug, the blood pressure showed a clear level change. Too many patients either showed no level change or their blood pressure even increased. We inspected the patient records that formed negative evidence against the rule. There were patients with arhythmic heart beat, patients where the blood pressure varied more than usual, patients where the rate of increasing blood pressure decreased but not the blood pressure itself. On one hand, the hypothesis can be made more precise: the drug decreases the rate of increasing blood pressure. On the other hand, the variance of blood pressure and its relation to arythmic heart beat forms a local pattern. Again, the user has to decide whether the data are wrong or the instances allow us to form an interesting local pattern. Hence, another reason for a set of instances being interesting is that they are negative evidence for a valid rule.

Without the claim of completeness, we may state that there are three reasons for a set of instances to be interesting:

- exceptions to an accepted general rule
- examples that are not covered by any valid rule
- negative examples that prevent the acceptance of a rule.

The acceptance of a rule hypothesis depends on the acceptance criterion or quality measure. In the preceding paragraphs we argued on the basis of a general quality measure which combines positive and negative evidence for and against the validity of a rule in some manner. Such a measure is tailored to finding general models. The pattern detection then means to find the deviations from the general rules. There is an alternative, though. The quality measure can be tailored to finding interesting sets of instances directly. Subgroup detection is the learning task of finding parts of the overall instance space, where the probability distribution differs significantly from the global one [12]. The subgroups can be considered local patterns which are selected by a quality criterion that implements the refutation of the null hypothesis, namely that the distribution in the subgroup equals that in the entire population. A logical approach to subgroup detection has been developed by Stefan Wrobel [18]. The same subgroups can be found by the collection of negative or uncovered examples and by a well-suited quality measure for subgroup detection. Hence, in principle the two approaches fall into the same category. The algorithms, however, are different. Subgroup detection explores the overall search space and uses pruning methods and language restrictions in order to become tractable. The two-step approach allows to use different search spaces and learning methods for the general model and the local patterns. Since the number of negative or uncovered examples is comparatively small, a larger search space can be explored by a more demanding procedure.

In this paper, we describe an inductive logic programming algorithm which first finds all valid rules in a very restricted search space. These rules are used to collect interesting instances as described above. The second learning step then finds definitions for the small samples in an enlarged search space. There are three ways to enlarge the hypothesis space for the learning algorithm:

- the synactic language restriction of hypotheses is weakened
- the dimensionality of the examples is increased, i.e. more (possibly finer grained) attributes are taken into account
- a more expensive learning strategy is chosen

We investigate these three options on the different sets of interesting instances. For illustration we use the movie database which is freely available in the internet (www.imdb.com). The paper is organised as follows. First, we describe the learning algorithm RDT/DM and indicate the size of the hypothesis space depending on the syntactic language restrictions, the dimensionality of examples and the learning strategy. Second, we shortly present results of learning global models (i.e. all valid rules) and discuss some quality measures for accepting hypotheses. Third, the sets of interesting instances and the detection of local patterns is illustrated by our movie application. A discussion relating the approach to others concludes the paper.

2   Inductive Logic Programming Using RDT/dm

The rule learning task has been stated within the inductive logic programming (ILP) paradigm by Nicolas Helft [9] using the notion from logic of minimal models of a theory $\mathcal{M}^+(Th) \subseteq \mathcal{M}(Th)$. Of course, in general there may well exist many minimal models. However, for first-order logic there exist restrictions such that there is exactly one minimal model to a theory.

Definition 1. *(Minimal model) An interpretation I is a model of a theory Th, $\mathcal{M}(Th)$, if it is true for every sentence in Th. An interpretation I is a minimal model of Th, written $\mathcal{M}^+(Th)$, if I is a model of Th and there does not exist an interpretation I' that is a model of Th and $I' \subset I$.*

Rule Learning
Given observations $\mathcal{E}$ in a representation language $\mathcal{L}_\mathcal{E}$ and background knowledge $\mathcal{B}$ in a representation language $\mathcal{L}_\mathcal{B}$,
find the set of hypotheses $\mathcal{H}$ in $\mathcal{L}_\mathcal{H}$, which is a (restricted) first-order logic, such that

(1) $\mathcal{M}^+(\mathcal{B} \cup \mathcal{E}) \subseteq \mathcal{M}(H)$ (validity of H)
(2) for each $h \in \mathcal{H}$ there exists $e \in \mathcal{E}$ such that $\mathcal{B}, \mathcal{E} - \{e\} \not\models e$ and $\mathcal{B}, \mathcal{E} - \{e\}, h \models e$ (necessity of h)
(3) for each $h \in \mathcal{L}_\mathcal{H}$ satisfying (1) and (2), it is true that $\mathcal{H} \models h$ (completeness of $\mathcal{H}$)
(4) There is no proper subset $\mathcal{G}$ of $\mathcal{H}$ which is valid and complete (minimality of $\mathcal{H}$).

The first sentence states that the minimal model of the observations and the background knowledge must be a subset of the model of the learned rules. There is one interpretation for learning result, examples, and background knowledge. This sentence expresses the *correctness* of the learning result. The other sentences approximate the completeness and minimality of theories. Since Kleene's proof that we cannot find the minimal set of axioms to a set of observations (facts), the minimality had to be reduced to not including rules that could be removed without any loss. It is still possible that by combining some rules into one rule, the set of rules that is valid and complete is smaller than the learning result. We cannot escape this possible shortcoming of too large a learning result. What we can escape is having redundant rules in the learning set that do not contribute to covering observations. This is stated in (2) and (4). The third sentence states that all rules that are valid and necessary in the sense of (2) are included in the learning result. Again, this property does not exclude that there are more elegant and concise rules that are not learned, since this would contradict Kleene's proof. It only states that all correct rules that are necessary for covering examples are included in the learning result. This learning task has been taken up by several ILP researchers, e.g., [10], [5], [3]. Rule learning is more difficult than the concept learning task because it is possible that all results of concept learning could also be found by rule learning, but not vice versa [11]. Since the learning task is hard in terms of computational complexity, it must be constrained. Constraining the hypothesis language $\mathcal{L_H}$ to less expressive power than first-order logic is the key to making rule learning efficient. The clear learnability border has been shown for restrictions of first-order logic [11].

In order to restrict the hypothesis space, RDT/DM uses a declarative specification of the hypothesis language, just as its predecessor RDT does [10]. The specification is given by the user in terms of rule schemata. A rule schema is a rule with predicate variables (instead of predicate symbols). In addition, arguments of the literals can be designated for learning constant values. Examples of rule schemata are:

$$mp1(P1, P2, P3) : P1(X1)\&P2(X1) \rightarrow P3(X2)$$
$$mp2(P1, P2, P3) : P1(X1, X2)\&P2(X2) \rightarrow P3(X1)$$
$$mp3(P1, P2, P3, P4) : P1(X1)\&P2(X1)\&P3(X1) \rightarrow P4(X1)$$
$$mp4(C, P1, P2, P3) : P1(X1, C)\&P2(X1) \rightarrow (P3(X1)$$

Where the first and third rule schema restricts $\mathcal{L_H}$ to propositional learning, the second and fourth schema expresses a simple relation. In the last schema, the second argument of the first literal is a particular constant value that is to be learned. This is indicated in the meta-predicate by C.

For hypothesis generation, RDT/DB instantiates the predicate variables and the arguments that are marked for constant learning. The arity of a predicate as well as the sorts of arguments are taken into account, so that only sort-correct and fitting predicates instantiate the predicate variables. A fully instantiated rule schema is a rule. An instantiation is, for instance,

mp1(america, drama,top) $america(X1)\&drama(X1)\& \rightarrow top(X1)$

The rule states that a movie that was produced at the American continent and is of genre "drama" belongs to the top hundred of the movie rating as given by the movie database.

The rule schemata are ordered by generality: for every instantiation of a more general rule schema there exist more special rules as instantiations of a more special rule schema, if the more special rule schema can be instantiated at all. Hence, the ordering of rule schemata reflects the generality ordering of sets of rules. This structure of the hypothesis space is used while doing top-down search for learning. If a rule is learned its specialization w.r.t. the generality ordering of rule schmeata will not be tried, since this would result in redundant rules. Hence, RDT/DM delivers most general rules. The user writes the rule schemata in order to restrict the hypothesis space. The user also supplies a list of the predicates that can instantiate predicate variables (i.e. determines the dimensionality of the examples). This list can be a selection of all predicates in $\mathcal{L}_\mathcal{E}$.

Another kind of user-given control knowledge is the acceptance criterion. It is used to test hypotheses. The user composes an acceptance criterion for a rule $premise \rightarrow conclusion$ out of four terms which count frequencies in the given data:

$pos(h)$ the number of supporting instances: $fr(premise \wedge conclusion)$;
$neg(h)$ the number of contradicting instances: $fr(premise \wedge \neg conclusion)$;
$concl(h)$ the number of all tuples for which the conclusion predicate of the hypothesis holds: $fr(conclusion)$; and
$negconcl(h)$ the number of all instances for which the conclusion predicate does not hold: $fr(\neg conclusion)$.

This general form for criteria directly corresponds to interestingness criteria as presented in [8]. Using the four terms, one can easily express different acceptance criteria. A typical one (similar to that of APRIORI [1]) is:

$$\frac{pos(h)}{concl(h)} - \frac{neg(h)}{concl(h)} \geq 0.8$$

The acceptance criterion can also be written in a Bayesian manner. If there are two classes for the conclusion predicate (e.g., top 100 movies and bottom 100 movies), the following criterion is similar to the requirement that the a posteriori probability must equal or exceed the a priori probability:

$$\frac{pos(h)}{pos(h) + neg(h)} \geq \frac{concl}{concl + negconcl}$$

RDT/DM is similar to RDT/DB [2] in that it directly accesses the ORACLE database system. It is a re-implementation using the JAVA programming language. The main difference is the search stratey in the hypothesis space $\mathcal{L}_\mathcal{H}$. Where RDT/DB performs a breadth-first search which allows safe pruning, RDT/DM performs a depth-first search in order to minimize the database accesses and to exploit already selected subsets of records. For each accepted hypothesis h its instances $ext(h)$ are collected as $pos(h)$ (support) and $neg(h)$

(outliers). The uncovered instances of all rules $S - ext(H)$ are also stored[1]. The data dictionary of the database system contains information about relations and attributes of the database. This information is used in order to map database relations and attributes automatically to predicates of RDT's hypothesis language. For hypothesis testing, SQL queries are generated by the learning tool and are sent to the database system. The counts for $pos(h), neg(h), concl(h)$, and $negconcl(h)$ are used for calculating the acceptance criterion for fully instantiated rule schemata.

The size of the hypothesis space of RDT/DM does not depend on the number of database records, but on the number of rule schemata, r, the number of predicates that are available for instantiations, p, and the maximal number of literals of a rule schema, k. For each literal, all predicates have to be tried. Without constant learning, the number of hypotheses is $r \cdot p^k$ in the worst case. As k is usually a small number in order to obtain understandable results, this polynomial is acceptable. Constants to be learned are very similar to predicates. For each argument marked for constant learning, all possible values of the argument (the respective database attribute) must be tried. We write c for the number of constants that are marked for learning in a rule schema. Let i be the maximal number of possible values of an argument marked for constant learning; then, the hypothesis space is limited by $r \cdot (p \cdot i^c)^k$. The size of the hypothesis space determines the cost of hypothesis generation. For each hypothesis, two SQL statements have to be executed by the database system. These determine the cost of hypothesis testing.

2.1   The Movie Database and the Chosen Hypothesis Spaces

The movie database imdb.com stores millions of movies with their genre, actors, producers, directors, the production country, keywords, and year of publication. For actors and directors tables with further information about them exist. What makes the database of interest to us is the voting of movie visitors and the resulting ranking of movies into the top 100 movies and the bottom 100 movies. For our experiments we selected only the top and bottom 100 movies. The first task was to learn all valid rules about the top 100 movies. Hence, we reduced the first learning task more or less to concept learning, here, the classification of top movies. This exercise in automatic modeling or learning global models is meant to be the basis for detecting local patterns in the second step. Hence, the restriction should not be a problem.

In order to map database relations and attributes to predicates, the system offers a tool which constructs predicates using the data dictionary of the database and exploiting foreign key relations. In our experiments we used two mappings.

Mapping 1: For each relation R with attributes $A_1, \ldots, A_n$, where the attributes $A_j, \ldots, A_l$ are the primary key, for each $x \in [1, \ldots, n] \backslash [j, \ldots, l]$ a predicate $rn\_AX(A_j, \ldots, A_l, A_x)$ is formed, where AX is the string of the attribute name.

---

[1] S is the set of all instances.

Since the primary key of the relations is a single attribute, we get two–place predicates. The number of predicates is bound by the number of relations times the maximal number of attributes of a relation (without key attributes).

The other mapping reduces the expressiveness to propositional logic. Of course, this means that the size of the hypothesis space is reduced.

Mapping 2: For each attribute A_i which is not a primary key and has the values $a_1, \ldots, a_n$ a set of predicates $rn\_AI\_ai(A_j, \ldots, A_l)$ are formed, $A_j, \ldots, A_l$ being the primary key.

Mapping and rule schemata together determine the size of the hypothesis space. For learning the general model of top movies, we used the first two rule schemata shown above and the first and second mapping. 22 predicates were selected for learning. In the propositional case this corresponds to 22 dimensions of movies. In the worst case only $2 \cdot 22^2 = 968$ hypotheses need to be tested[2]. For the detection of local patterns in selected instance sets, we enlarged the hypothesis space by including the third rule schemata shown above and by forming more predicates. Changing the learning strategy to constant learning using the fourth rule schema further enlarges the hypothesis space.

3   Learning the Global Model for Movie Ranking

Given the top 100 movies classified $top(MovieID)$ and the bottom 100 movies classified $not(top(MovieID))$, we instantiated the conclusion predicate of the first three meta-predicates by $top(X)$. Each movie was characterised by predicates with arity 1 for genre and production country or continent and by predicates with arity 2 for the relation between a movie and any actor of it as well as the relation between a movie and its director. Actors and directors were described by a one–place predicate stating, whether the actor performed in or the director made at least 2 of the top 100 movies, namely $topActor(X)$ and $topDir(X)$. Similarly, $noBotActor(X)$ respectively $noBotDir(X)$ states that an actor or director was never involved in one of the bottom movies. The acceptance criterion was set to

$$\frac{pos(H)}{concl(H)} - \frac{neg(H)}{concl(H)} \geq 0.3$$

These are the learned rules:

$h_1 : usa(X)\&drama(X) \rightarrow top(X)$
$h_2 : director(X,Y)\&topDir(Y) \rightarrow top(X)$
$h_3 : actor(X,Y)\&topActor(Y) \rightarrow top(X)$
$h_4 : director(X,Y)\&noBotDir(Y) \rightarrow top(X)$
$h_5 : actor(X,Y)\&noBotactor(Y) \rightarrow top(X)$

---

[2] Since the conclusion predicate is always instantiated by $top(X)$, the conclusion literal is not counted.

Of course, the rules do not make sense for cineasts, but only for the ranking of the movie database in the excerpt we used. The ranking clearly favours American movies: 68 of the top movies and 82 of the bottom movies are American. Taking the genre into account, 40 of the top movies are American dramas, but only 7 of the bottom movies. The tolerant acceptance criterion allows us to collect 7 outliers. 15 top movies are uncovered by the rules. These are candiates for local patterns hidden by the overwhelming dominance of American movies.

We wonder whether another acceptance criterion would allow us to describe the uncovered instances already in the first step. It should be a criterion which focuses on subgroups within the data. Hence, we applied the criterion based on the Binomial test heuristic:

$$\sqrt{\frac{pos(h)+neg(h)}{concl+negconcl}} \cdot \mid \frac{pos(h)}{pos(h)+neg(h)} - \frac{concl}{concl+negconcl} \mid \geq 0,05$$

The first factor weights the size of $ext(h)$ in relation to the samples size and the second factor compares the distribution in $ext(h)$ with that of the overall population. A theoretical investigation of quality functions as this one and its distinction from averaging functions can be found in [16]. Here, we identify the data set with the overall population and consider $ext(h)$ a sample. The significance of a difference in the distribution is determined with respect to the null hypothesis. In fact, additional rules are found that cover some of the previously uncovered instances:

$h_6 : italy(X)\&drama(X) \rightarrow top(X)$
$h_7 : denmark(X)\&drama(X) \rightarrow top(X)$

However, many more rules were also found and cover instances redundantly. If we narrow the pruning criterion, we get exactly the rules learned in the first step. Hence, we either receive too many rules, among them senseless ones, or we end up with the global rules.

4   Interesting Instances in the Movie Database

Our aim is to detect local patterns. We have learned general rules in order to get rid of the dominant process. We now want to inspect the remaining instances using a larger hypothesis space. We used additionally the third rule schema and changed the acceptance criterion to

$$\frac{pos(h)}{pos(h)+neg(h)} - \frac{neg(h)}{pos(h)+neg(h)}$$

We enlarged the number of predicates for learning. The table "keywords" contains up to 200 words for each movie. We formed 286 one-ary predicates out of these. That is, for the few interesting instances we offered a hypothesis space of a hundred million hypotheses ($3 \cdot 330^3 = 107811000$).

The first potentially interesting set of instances are the *outliers* to accepted rules. Do these American dramas which are ranked within the bottom movies have something in common? We switched the learning strategy to learning constant values for actors and directors. The possible values for the argument marked for constant learning are 2544. Hence, the hypothesis space consists of about 10^{17} hypotheses, $4 \cdot (330 \cdot 2544^1)^3$. 6 rules were found which characterize bottom movies covered by general rules for top movies. Three rules blamed certain actors which performed in none of the top movies but only in the bottom ones[3]. Another rule found the combination of "drama" and "musical", two other rules refer to directors who made bottom movies or top movies. The rules covered only about half of the exceptions. Decreasing the threshhold value of the acceptance criterion led to 34 actors who play in bottom movies and to some keywords for American bottom films, among them "police", "screwball", and "independent". However, more rules were learned than there are exceptions. Decreasing the selectivity of the criterion leads to arbitrary rules. Hence, using Occam's razor, we prefer the 6 rules and list the remaining instances.

The second possibly interesting set of instances are the *uncovered examples*. With the enlarged set of rule schemata but without the additional predicates, h_6 and h_7 were found, indicating the subgroups of Italian and Danish dramas. A third learned rule states that those European movies are top, which are classified as "drama" and as "romance" as well. However, these rules still leave some instances uncovered. Using the enlarged set of predicates 7 rules were found. Rule h_6 covering particularly the famous movies "bicycle thieves" (1948) and "La vita e bella"(1997) is again found. The keywords "independent film" or "family" make a European movie top. Independent films such as the Danish dogma movies that led to rule h_7 are covered. The keyword "love" together with the genre "drama" also characterise the uncovered instances. Funny enough, the keyword "bicycle" is also characteristic for European top movies as well as for those of genre "drama". All rules do make sense and it is not surprising that 6 of the 7 rules deal with European movies. These seem to be the ones that are dominated by the American movie population[4].

The third collected set of instances was $neg(h)$ where h was rejected in the first learning step. In our movie application, this set of instances was quite large (85) and uninteresting. It shows that the general model is quite appropriate and learning the global model excludes successfully senseless rules. No further learning was performed on this sample.

5   Conclusion

Instance selection usually means to reduce the original data set such that the reduced set allows to find the general model underlying the original data. Sampling is its most popular technique, search for critical points or the construction of prototypes are others [13].The detection of outliers in such a context aims at easing learning from the reduced set [6].In this paper, we are concerned with

---

[3] What the learning algorithm did not find because we did not apply constant learning to movie titles, is that these actors play in "Police Academy"...

[4] It is a pity that we could not find the nationalities of the voting population in the database, because it could well be that the aubgroups actually do not refer to the producing nation of the movies but to the nation of the voters.

the opposite goal of finding instance sets that allow us to find local patterns that contrast the general model. The instance sets do not mirror the overall distribution as do (random) samples. The detection of outliers in this context means to focus on small abnormal subsets of the data. Similarly, the collection of uncovered instances focuses on subsets which do not fit the general model. These subsets of data are candidates for being the extension of local patterns. We consider local patterns the effect of behavior of a distinct, small population that is dominated by the large population. In other words, we assume a mixture distribution made of distinct components, where one is overwhelmingly large. Therefore, usual methods like, e.g., Markov switching models or models of finite mixture distributions do either not detect these small groups or detect overly many small groups.

The two-step approach presented here, excludes the large population in its first step and hence prevents us from finding too many subgroups. In spirit, our first experiment on outliers is similar to the demand-driven concept formation of Stefan Wrobel [17]. In an incremental setting, he collected the exceptions of strongly supported rules and learned their definition. The hypothesis space remained the same for rule learning and defining new concepts over exceptions. He did not deal with uncovered instances. In our movie application, the first step characterises American movies (the majority group of the population). The collection of uncovered positive examples worked particularly well. The second step then concentrates on European movies (the minority group of the population). From an algorithmic point of view, the two-step procedure allows us to change the search space and search method within a learning run. The first step processes the overall data set using a small hypothesis space, i.e. it estimates the overall distribution on the basis of low-dimensional instances and a tight restriction of hypotheses' complexity. The second step processes the selected abnormal instances on the basis of high-dimensional instances and more complex hypotheses. Here, we applied the same learning algorithm in both steps. Further work could apply complex algorithms in the second step such as the lgg [15] in order to bottom-up generalise interesting instances. We have illustrated our approach by the movie database with an excerpt of four tables. Further experiments and – more important – theoretical considerations are still needed.

Acknowledgment. Dirk Münstermann has developed RDT/DM and has run the experiments on the movie database. His diploma thesis which explains the new search strategy and data management is forthcoming.

References

1. Rakesh Agrawal, Heikki Mannila, Ramakrishnan Srikant, Hannu Toivonen, and A. Inkeri Verkamo. Fast discovery of association rules. In Usama M. Fayyad, Gregory Piatetsky-Shapiro, Padhraic Smyth, and Ramasamy Uthurusamy, editors, *Advances in Knowledge Discovery and Data Mining*, chapter 12, pages 307–328. AAAI Press/The MIT Press, Cambridge Massachusetts, London England, 1996.

2. Peter Brockhausen and Katharina Morik. Direct access of an ILP algorithm to a database management system. In Bernhard Pfaringer and Johannes Fürnkranz, editors, *Data Mining with Inductive Logic Programming (ILP for KDD)*, MLnet Sponsored Familiarization Workshop, pages 95–110, Bari, Italy, jul 1996.
3. L. DeRaedt and M. Bruynooghe. An overview of the interactive concept–learner and theory revisor CLINT. In Stephen Muggleton, editor, *Inductive Logic Programming.*, number 38 in The A.P.I.C. Series, chapter 8, pages 163–192. Academic Press, London [u.a.], 1992.
4. T. Fawcett and F. Provost. Adaptive fraud detection. *Data Mining and Knowledge Discovery*, 1(3):291 – 316, 1997.
5. P. A. Flach. A framework for inductive logic programming. In Stephen Muggleton, editor, *Inductive Logic Programming.*, number 38 in The A.P.I.C. Series, chapter 9, pages 193–212. Academic Press, London [u.a.], 1992.
6. Dragan Gamberger and Nada Lavrac. Filtering noisy instances and outliers. In Huan Liu and Hiroshi Motoda, editors, *Instance Selection and Construction for Data Mining*, pages 375 – 394. Kluwer, 2001.
7. Isabelle Guyon, Nada Matic, and Vladimir Vapnik. Discovering informative patterns and data cleaning. In Usama M. Fayyad, Gregory Piatetsky-Shapiro, Padhraic Smyth, and Ramasamy Uthurusamy, editors, *Advances in Knowledge Discovery and Data Mining*, chapter 2, pages 181–204. AAAI Press/The MIT Press, Menlo Park, California, 1996.
8. David Hand, Heikki Mannila, and Padhraic Smyth. *Principles of Data Mining*. Massachusetts Institute of Technology, 2001.
9. Nicolas Helft. Inductive generalisation: A logical framework. In *Procs. of the 2nd European Working Session on Learning*, 1987.
10. J.-U. Kietz and S. Wrobel. Controlling the complexity of learning in logic through syntactic and task–oriented models. In Stephen Muggleton, editor, *Inductive Logic Programming.*, number 38 in The A.P.I.C. Series, chapter 16, pages 335–360. Academic Press, London [u.a.], 1992.
11. Jörg Uwe Kietz. *Induktive Analyse relationaler Daten*. PhD thesis, Technische Universität Berlin, Berlin, oct 1996.
12. Willi Klösgen. *Handbook of Knowledge Discovery and Data Mining*, chapter Subgroup patterns. Oxford University Press, London, 2000. 2000 to appear.
13. Huan Liu and Hiroshi Motoda. *Instance Selection and Construction for Data Mining*. Kluwer Publishers, 2001.
14. Katharina Morik and Peter Brockhausen. A multistrategy approach to relational knowledge discovery in databases. *Machine Learning Journal*, 27(3):287–312, jun 1997.
15. Gordon D. Plotkin. A note on inductive generalization. In B. Meltzer and D. Michie, editors, *Machine Intelligence*, chapter 8, pages 153–163. American Elsevier, 1970.
16. Tobias Scheffer and Stefan Wrobel. A Sequential Sampling Algorithm for a General Class of Utility Criteria. In *Proceedings of the International Conference on Knowledge Discovery and Data Mining*, 2000.
17. Stefan Wrobel. *Concept Formation and Knowledge Revision*. Kluwer Academic Publishers, Dordrecht, 1994.
18. Stefan Wrobel. An algorithm for multi–relational discovery of subgroups. In J. Komorowski and J. Zytkow, editors, *Principles of Data Minig and Knowledge Discovery: First European Symposium (PKDD 97)*, pages 78–87, Berlin, New York, 1997. Springer.

Complex Data: Mining Using Patterns

Arno Siebes[1] and Zbyszek Struzik[2]

[1] Utrecht University
Utrecht, The Netherlands
arno@cs.uu.nl
[2] CWI
Amsterdam, The Netherlands
zbyszek@cwi.nl

Abstract. There is a growing need to analyse sets of complex data, i.e., data in which the individual data items are (semi-) structured collections of data themselves, such as sets of time-series. To perform such analysis, one has to redefine familiar notions such as similarity on such complex data types. One can do that either on the data items directly, or indirectly, based on features or patterns computed from the individual data items. In this paper, we argue that wavelet decomposition is a general tool for the latter approach.

1 Introduction

One of the many variants of Moore's law [10] is the exponential growth of hard-disk capacity per euro. This growth has enabled the rise of a fast increasing number of fast growing *complex data* sets. That is, data collections in which the individual data items are no longer "simple" (atomic in database terminology) values but are (semi-)structured collections of data themselves. For example, large text corpera, multi-media data collections, databases with milions of time-series, and large collections of DNA and/or protein data.

With the advent of sets of complex data grows the need to analyse such collections. This is, e.g., illustrated by the existence of satelite workshops on topics such as multimedia data mining, text mining, and spatio-temporal data mining around all the major KDD conferences.

To apply standard or new data-analysis techniques, fundamental notions that or obvious for numbers have to be redefined. One example of such a notion is *similarity*, which is essential, e.g., for clustering and classification [5]. There are two ways in which one can define similarity for complex data types. Firstly, one can define it directly on the complex data items. Secondly, one derive features or patterns from complex data items and define the similarity on these features or patterns.

Both approaches have their respective advantages. The former allows the definition of, e.g., a similarity measure that is based on all aspects of the data items rather than only on those that are represented by features. Measures that find their roots (ultimately) in *Kolmogorov complexity* [8] are an example of this

D.J. Hand et al. (Eds.): Pattern Detection and Discovery, LNAI 2447, pp. 24–35, 2002.
© Springer-Verlag Berlin Heidelberg 2002

class. The latter, in contrast, allow one to take only relevant "local structure" into account for the similarity measure, such as measures based on *wavelet decomposition* [12]. Which of the two is best, thus depends primarily on the notion of similarity one wants to use.

For both approaches, it is important that the way one defines the similarity measure is sufficiently general. That is, it is applicable across specific problems and across various complex data types. Otherwise the analysis of complex data would degenerate into a ragbag of ad-hoc techniques. For the first approach, Kolmogorov complexity seems a good candidate for such a general technique. We briefly discuss this in Section 2.

The main argument of this paper is that wavelets provide such a general technique for similarity (and other notions) based on local structure in the complex data items, this is discussed in the rest of the paper. In Section 3, we discuss how we have used wavelets on two problems on sets of time series. Since wavelets are not yet a standard tool in the data miners toolbox, this section also contains a brief introduction into wavelet decomposition. In Section 4, we argue that the wavelet decomposition is also applicable for other types of complex data. In the final section we formulate our conclusions.

Perhaps the simplest form of complex data is a collection of texts. For example, a set of articles or books. A recurrent problem in history has been: can we assign an author to a text. Either because the author has stayed anonymous or because it is suspected that the author has used a pseudonym. A controversial example of the latter is the question whether all of the works attributed to Shakespeare were actually written by Shakespeare.

A less controversial and perhaps more famous example in the statistical literature is that of the *The Federalist* papers [11]. These papers were written by Alexander Hamilton, John Jay, and James Madison to persuade the citizens of the State of New York to ratify the Constitution [11]. The author of most of these papers is known, but the authorship of 12 of them is in dispute between Hamilton and Madison. In their study [11] Mosteller and Wallace search for words whose usage-frequency discriminates between Hamilton and Madison.

Word-frequencies are probably the most simple features one can derive from text. But they play an important role in information retrieval and in text mining. For classification, however, word-frequencies are perhaps too simple features since other stylistic aspects are not taken into account at all. Hence, it is interesting to see what similarity measures one can define on the complete texts directly.

In a recent paper [1], the authors propose to use WinZip for the task of assigning papers to authors. The motivation is that compression algorithms such as Lempel-Ziv encoding exploit the structure in a text. Hence, for texts A and B, they define:

- L_A = the length of Lempel-Ziv encoding of A in bits,
- let b be a (small) introductory part of B:

$$\Delta_{Ab} = L_{A+b} - L_A$$

To test how well Δ_{Ab} can be used to assign texts to authors, they use a corpus of 90 texts by 11 authors and perform 90 experiments. In each experiment, they take one of the texts as produced by an unknown author X. To discover the author, the search for the (remaining) text A_i that minimizes $\Delta_{A_i x}$. The author of A_i is then predicted to be the author of X. In 84 out of the 90 experiments this prediction was correct.

On request of a Dutch newspaper, NRC Handelsblad, Benedetto et al have used this technique to solve an open case in contemporary Dutch literature: are the authors Arnon Grunberg and Marek van der Jagt one and the same person? Based on style-similarities and the elusiveness of the second author (only contact via e-mail), this is what critics suspected. Despite this controversy, van der Jagt got a prize for the best debut a couple of years after Grunberg picked it up.

Using a collection of other contemporary writers, the results pointed to Grunberg. Clearly, in itself this proves nothing. However, the author conceded. Fortunately, right after this show-down a new unknown writer, Aristide von Bienefelt, debuted. Again it is rumored that this is actually Grunberg.

To motivate the suitability of their approach, Benedetto et al refer to Kolmogorov complexity [8], but they do not actually explore this avenue. In that book, Li and Vitányi already introduced an *information distance* between two binary strings. In subsequent work, together with co-authors, they refined this notion, see e.g., [2] and showed that it applicable in many cases.

The Kolmogorov complexity of string x, denoted by $K(x)$ is the length of the shortest program of a universal computer that outputs x. Up to an additive constant, $K(x)$ is independent of the particular machine chosen. Similarly, $K(x|y^*)$ is defined as the length of the shortest program that computes x given y^* as input; see [8] for more details.

In [7] the normalised information distance between two strings x and y is defined by:

$$d(x,y) = \frac{max\{K(x|y^*), K(y|x^*)\}}{max\{K(x), K(y)\}}$$

Moreover, it is proved that this measure is universal in the sense that it minorizes every remotely computable type of dissimilarity measure.

The problem in applying such distance measures is that $K(x)$ is not computable. Hence, the trick in the applications is to use a suitable approximation of $K(x)$, see the cited literature for more details.

2   Complex Data: Time Series

Kolmogorov complexity is universal. However, there are problems and complex data types for which it is not obviously the best solution. For example, in cases where it is not the complete complex data item that is important, but "only" local patterns in that data. In this section, we discuss such an example and illustrate how we have tackled it using wavelets.

2.1   The Problem

Banks follow large collections (over a million items) of financial and economic time series, on stock-prices, exchange rates, interest rates, unemployment rates et cetera. This information is, e.g., used to compute the combined risk of the banks portfolio or to determine opportunities for stock-exchange transactions.

Unfortunately, the time-series are not without error. Since the data are used almost instantaneously and potentially large amounts of money are involved, it is of utmost importance that such errors are recognized as early as possible. The sheer number of time series implies that most of the work has to be done automatically, involving people only if one is fairly certain of a mistake. Both the number of false positives and the number of false negatives should be as small as possible. Large mistakes are easily recognisable, hence we focus on relatively small errors.

One way to attack this problem is by modelling each individual time-series and signal whenever the new reported value differs from the predicted value. The problem here lies obviously in the accuracy of the prediction. Although there are numerous studies in which the authors attempt to model stock prices, we have chosen another approach. If only because the number of data miners we know that live on an estate in the country and are driven around in Rolls-Royces is fairly low[1].

Our approach is based on the *behaviour* of time series.

Outliers: are signalled if an individual time-series suddenly behaves completely differently from its (recent) past.

Group-outliers: are time-series that suddenly behave differently from their *peer group*. These peer-groups are constructed by clustering time-series on their recent behaviour; note that the idea of peer-groups has also been used for fraud-detection in [3].

Note that in both cases, the similarity is based on the recent past. It is not a priori clear what time-frame defines the recent past. In other words, applying Kolomogorov complexity based techniques is far from straightforward. We have opted to characterise the recent past using wavelet decomposition. The advantage of such a multi-scale method is that one discovers what constitutes the recent past while analysing.

2.2   Introducing Wavelets

Since wavelets are not yet a common tool in data mining, we first give a brief (and simplified) introduction to wavelets. It is well-known that a function f supported on $[-s, s]$ can be represented by a *Fourier series*:

$$f(x) = \sum_{n=0}^{\infty} c_n e^{2\pi i n(x/2s)}$$

---

[1] Of course, it is entirely possible that these successful data miners do not publish their methods *because* they make so much money out of their ideas.

in which the Fourier transforms are computed by the convolutions with $e^{-2\pi it(n/2s)}$, i.e.,

$$c_n = \int_{-s}^{s} f(t)e^{-2\pi it(n/2s)}dt$$

Moreover, for reasonable functions f (integrable in the L^2 sense) this can be extended to the whole domain of f.

For our purposes, the disadvantage of global transformations such as the Fourier transform is that they do not support the local analysis of functions. That is, it is hard to see patterns The wavelet transforms provides such locality because of the limited (effective) support of wavelets. In addition, they poses the often very desirable ability of filtering the polynomial behaviour to some predefined degree. For time-series data this means that we can easily handle non-stationarities like global or local trends or biases. Last described, but certainly not least for our purposes, one of the main aspects of wavelet transforms is the ability to reveal the *hierarchy* of (singular) features including the scaling behaviour - the so-called *scale-free* behaviour.

Wavelet transforms are, like Fourier transforms, computed by convolutions. The difference is that we don't use a function of infinite support like e^x, but a function with a localised (effective) support: the wavelet.

Usually, one starts with a smoothing kernel θ. The wavelet is then a derivative ψ of the kernel θ. With Fourier transforms, we only have a scaling parameter s, but since wavelets have a limited domain, we also have a *translation parameter b*. That is, we compute transforms by convolutions with ψ. Denoting the convolution by $< f, \psi >$ we have:

$$Wf(s,b) = < f, \psi > (s,b) = \frac{1}{s}\int_{\Omega} f(x)\psi(\frac{x-b}{s})dx$$

In the continuous case, the the Gaussian smoothing kernel $\theta(x) = exp(-x^2/2)$ has optimal localisation in both frequency and position. An often used wavelet derived from this kernel is its second derivative which is called the *Mexican hat*[2].

If we compute the continuous wavelet transform (CWT) of fractional Brownian motion using the Mexican hat wavelet, we get figure 1.

The front axis is the position, the scale axis pointing "in depth" is (as traditional) in logarithmic scale, while the vertical axis denotes the magnitude of the transform. This 3D plot shows how the wavelet transform reveals more and more detail while going towards smaller scales, i.e. towards smaller $\log(s)$ values. This is why the wavelet transform is sometimes referred to as the 'mathematical microscope'.

As well as continuous wavelet transforms, there exist discrete transforms (DWT). In this case the simplest kernel is the block function:

$$\theta(x) = \begin{cases} 1 \text{ for } 0 \leq x \leq 1 \\ 0 \text{ otherwise} \end{cases}$$

---

[2] If you wonder why, draw its graph.

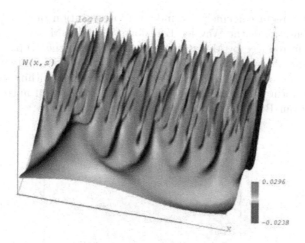

Fig. 1. CWT of Brownian motion

Which yields the *Haar* wavelet:

$$\psi(x) = \begin{cases} 1 & \text{if } 0 \leq x \leq 0.5 \\ -1 & \text{if } 0.5 \leq x \leq 1 \\ 0 & \text{otherwise} \end{cases}$$

For a particular choice of scaling and translation parameters, we get the Haar system:

$$\psi_{s,b}(x) = 2^{-s}\psi(2^{-s}x - b) \quad s > 0, b = 0, \ldots, 2^{s}$$

For an arbitrary time series $f = \{f_i\}_{i \in \{1,\ldots,2^N\}}$ on normalised support $[0,1]$ we have in analogy with the Fourier series:

$$f = f^0 + \sum_{m=0}^{N} \sum_{l=0}^{2^m} c_{m,l}\psi_{m,l}$$

In which $f^0 = <f, \theta>$ and $c_{m,l} = <f, \psi_{m,l}>$ Moreover, the approximations f^j of the time series f with the smoothing kernel $\theta_{j,k}$ form a 'ladder' of multi-resolution approximations:

$$f^{j-1} = f^j + \sum_{k=0}^{2^j} <f, \psi_{j,k}> \psi_{j,k},$$

where $f^j = <f, \theta_{j,k}>$ and $\theta_{j,k} = 2^{-j}\theta(2^{-j}x - k)$.

It is thus possible to 'move' from one approximation level $j - 1$ to another level j by simply adding (subtracting for j to $j-1$ direction), the detail contained in the corresponding wavelet coefficients $c_{j,k}, k = 0 \ldots 2^j$.

The CWT is an extremely redundant representation of the data. A more usful representation is the Wavelet Transform Modulus Maxima (WTMM) representation, introduced by Mallat [9]. A maxima line in the 3D plot of the CWT is a line where the wavelet transform reaches local maximum (with respect to the position coordinate). Connecting such local maxima within the continuous wavelet transform 'landscape' gives rise to the entire tree of maxima lines; the WTTM. For our Brownian motion, we get figure 2

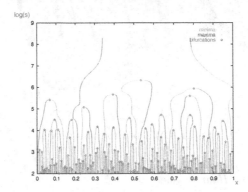

Fig. 2. WTTM plot of Brownian motion. The light lines are the minima lines, i.e., they connect the local minima, the dark lines the maxima lines, and the dots are the bifurcations

Restricting oneself to the collection of such maxima lines provides a particularly useful representation of the entire CWT. In particular (under suitable conditions), we have the following power law proportionality for the wavelet transform of the *isolated* cusp singularity in $f(x_0)$:

$$W^{(n)} f(s, x_0) \sim |s|^{h(x_0)} .$$

The exponents $h(x_0)$ are the local *Hölder exponents*, which describe the local *roughness* of the time-series The Hölder exponents provide a useful characterisation of the time-series as we will see in the next section.

The most direct representation of the time series with the Haar decomposition scheme would be encoding a certain predefined, highest, i.e. most coarse, resolution level s_{max}, say one year resolution, and the details at the lower scales: half (a year), quarter (of a year) etc., down to the minimal (finest) resolution of interest s_{min}, which would often be defined by the lowest sampling rate of the signals. The coefficients of the Haar decomposition between scales $s_{max}..s_{min}$ will be used for the representation:

$$Haar(f) = \{c_{i,j} : i = s_{max}..s_{min}, j = 1..2^i \} .$$

The Haar representation is directly suitable to serve for comparison purposes when the absolute (i.e. not relative) values of the time series (and the local slope) are relevant. In many applications one would, however, rather work with value independent, scale invariant representations. For that purpose, we have used a number of different, special representations derived from the Haar decomposition WT. To begin with, we will use the sign based representation. It uses only the sign of the wavelet coefficient:

$$s_{i,j} = sgn(c_{i,j})$$

Another possibility to arrive at a scale invariant representation is to use the difference of the logarithms (DOL) of values of the wavelet coefficient at the highest scale and at the working scale:

$$v_{i,j}^{DOL} = log(|c_{i,j}|) - log(|c_{1,1}|) ,$$

where i, j are working scale and position respectively, and $c_{1,1}$ is the first coefficient of the corresponding Haar representation. Note that the sign representation $s_{i,j}$ of the time series is complementary/orthogonal to the DOL representation.

The DOL representation can be conveniently normalised to give the rate of increase of v^{DOL} with scale:

$$h_{i,j} = v_{i,j}^{DOL} / log(2^{(i)}) \quad \text{for } i > 0 .$$

This representation resembles the Hölder exponent approximation introduced above at the particular scale of resolution

2.3   Mining with Wavelets

To illustrate the usefulness of wavelets for our time-series problems, we recall some of our published results. We start by looking for outliers in the single time-series of figure 3.

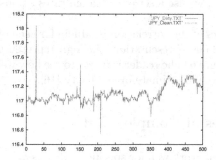

Fig. 3. The original time-series including spikes

The red spikes are obvious and are removed before we start our analysis. The local Hölder exponents are plotted in figure 4. By thresholding on h we separate the outliers from the rest. The reader is referred to [15] for more details.

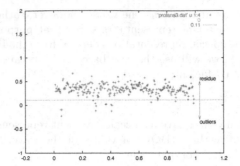

Fig. 4. The local Hölder exponents of our time-series

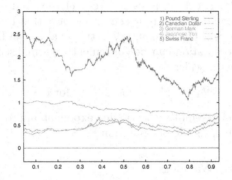

Fig. 5. The exchange rate of 1: Pound Sterling, 2: Canadian Dollar, 3: German Mark, 4: Japanese Yen, and 5: Swiss Frank against the US Dollar

For the similarity, we use a set of exchange rates against the dollar depicted in fig 5

We plot the values of the correlation products for each of the pairs compared, obtained with the Haar representation, the sign representation and the Hölder representation in figure 6. The reader is urged to see how well the representations match his/her own visual estimate of similarity [16].

3   Other Types of Complex Data

The fact that wavelets work well for specific problems on sets of time-series does not prove their generality. To argue that indeed they do form a widely applicable tool, we briefly discuss two other complex data types in this section. Firstly we discuss multimedia data, secondly we focus on DNA data. Since wavelets originate in signal analysis, it should not come as a surprise that they are useful for multimedia data, which is after all a collection of signals. That DNA data is also a possible application area is perhaps more surprising.

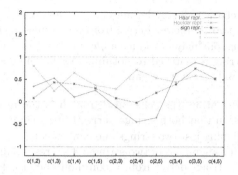

Fig. 6. The correlations for each of the pairs for the Haar, sign, and Hölder representation. Note, c(2, 5) stands for the correlation of the pair Canadian Dollar - Swiss Frank.

3.1 Multimedia Data

Consider for example a database with tens or hundreds of thousands or even millions of pictures. Ideally, each of this pictures would be amply annotated such that one could find a picture of Marilyn Monroe in the sunset near the Golden Gate bridge by simply typing in a few keywords. However, annotating such large collections is prohibitively expensive. In fact, it is wishful thinking that one could get coherent annotation schemes for pictures in a loosely coupled distributed database as the world wide web.

Therefore, a large part of research in multi-media databases is devoted to exploitation of patterns in the data items to enable searching for similar items in such databases. This would allow, e.g., searching for pictures that are similar to a given picture and, more difficult, to search for pictures using a sketch.

The difficulty of the problem is perhaps best illustrated by the wide variety of of techniques persued for image databases. They range from global colour histograms in some appropriate colour-space via collections of local colour histograms and textures to wavelet transforms [13]. With the possible exception of global colour histograms, all these techniques focus on (local) structures in the images to define similarity measures. The success of wavelet measures of similarity for retrieval point to the possible usefulness of such measures for data mining; see [14,17] for such and other appraoches.

3.2 DNA Data

The difference between species and between individual of one species is visible at the molecular level as smaller or larger differences in their DNA. Because of the mechanisms of evolution, the DNA of related species is highly similar or *homologuous*. *Alignment* of two DNA strings is (in-exact) string matching while maximising homology. The string matching for DNA differs from exact matching in two ways [4].

1. Two different characters may be matched for a cost that reflects the probability that the two characters derive from a common ancestor.
2. While matching, one may introduce a *gap* in one of the two strings; i.e., part of the other string is matched to nothing. Again, there are costs associated with starting and extending a gap.

With the cost information, the "cheapest" match is the most likely alignment under the assumption that both strings derive from a common ancestor.

Rather than aligning just two strings one can also align multiple strings using the same costs as for alignment of two strings. This results in an alignment that is most likely if all strings derive from a common ancestor [4]. Aligning two or more strings can, e,g, be used to cluster genomes to recover the path of evolution in phylogenetic analysis [4].

For many problems, using the whole, or large parts of the, genome is best. However, there are problems where one is only interested in local homology. One such example is caused by the fact that bacteria are able to swap genes across species. In such cases, local patterns can be useful.

Given the discrete nature of DNA data, it is perhaps surprising that wavelet transforms can be exploited here. However, in [6] the authors show that one can built an index on DNA strings using wavelets that facilitates range and k-nearest neighbour searches.

They define the frequency vector $f(s)$ of a string s as the vector $[n_A, n_C, n_T, n_G]$, in which n_X denotes the frequency of X in s. One vector for each DNA string in the database is clearly insufficient. Hence, they built a a hierarchy of wavelet transforms of the string, where each transform is computed from the function f on substrings of s. Of course, on smaller scales smaller substrings are used. In other words, again the wavelet transforms focus on smaller and smaller details of the data item.

The paper shows that this index works well if the edit distance is used. That is, the *costs* we discussed earlier do not play a role. But the authors plan to extend their work in this direction.

4   Conclusions

There are more and more complex datasets that need to be analysed. Such an analysis requires that familiar notions such as similarity are generalised to complex data types. There are two ways in which one can approach this problem. In the first approach one defines the similarity directly on the complex data types. In the second approach, one first computes local features or patterns on the individual data items and defines the similarity on these local patterns.

In this paper we have argued that wavelets are a general tool for this second approach. More in particular, we have discussed how we have used wavelets to detect errors in collections of time-series. Moreover, we have discussed how wavelets are used for multimedia data and DNA data.

Clearly, this paper is neither the first nor the last word on this topic. We plan more applications of wavelets on complex data types to convince others and ourselves of their use in the data miners toolbox.

References

1. Dario Benedetto, Emanuele Caglioti, and Victor Loreto. Language trees and zipping. *Physical Review Letters*, 88(4), 2002.
2. C.H. Bennet, P. Gács, M. Li, P.M.B. Vitányi, and W. Zurek. Information distance. *IEEE Trans. on Information Theory*, 44(4):1407–1423, 1998.
3. R.J. Bolton and D.J. Hand. Unsupervised profiling methods for fraud detection. *Credit Scoring and Control VII, Edinburgh*, 2001.
4. R. Durbin, S. Eddy, A. Krogh, and G. Mitchison. *Biological Sequence Analysis – Probabilistic Models of Proteins and Nucleic Acids*. Cambridge University Press, 1998.
5. David Hand, Heikki Mannila, and Padhraic Smyth. *Principles of Data Mining*. MIT Press, 2001.
6. Tamer Kahveci and Ambuj K. Singh. An efficient index structure for string databases. In *Proceedings of the 27th VLDB*, pages 351–360. Morgan Kaufmann, 2001.
7. Ming Li, Xin Li, Bin Ma, and Paul Vitányi. *Normalized Information Distance and Whole Mitochondrial Genome Phylogeny Analysis*. arXiv:cs.CC/0111054v1, 2001.
8. Ming Li and Paul Vitányi. *An Introduction to Kolmogorov Complexity and its Applications*. Springer Verlag, 1993.
9. S.G. Mallat and W.I. Wang. Singularity detection and processing with wavelets. *IEEE Trans. on Information Theory*, 38, 1992.
10. Gordon E. Moore. Cramming more components onto integrated circuits. *Electronics*, 38(8), 1965.
11. Frederick Mosteller and David L. Wallace. *Applied Bayesian and Classical Inference – The Case of The Federalist Papers*. Springer Verlag, 1984.
12. R. Todd Ogden. *Essential Wavelets for Statistical Applications and Data Analysis*. Birkhäuser, 1997.
13. Simone Santini. *Exploratory Image Databases – Content-Based Retrieval*. Academic Press, 2001.
14. Simeon J. Simoff and Osmar R. Zaïane, editors. *Proceedings of the First International Workshop on Multimedia Data Mining,MDM/KDD2000*. http://www.cs.ualberta.ca/ zaiane/mdm_kdd2000/, 2000.
15. Zbigniew R. Struzik and Arno Siebes. Wavelet transform based multifractal formalism in outlier detection and localisation for financial time series. *Physica A: Statistical Mechanics and its Applications*, 309(3-4):388–402, 2002.
16. Z.R. Struzik and A.P.J.M. Siebes. The haar wavelet in the time series similarity paradigm. In *Proceedings of PKDD99, LNAI 1704*, pages 12–22. Springer Verlag, 1999.
17. Osmar R. Zaïane and Simeon J. Simoff, editors. *Proceedings of the Second International Workshop on Multimedia Data Mining, MDM/KDD2001*. http://www.acm.org/sigkdd/proceedings/mdmkdd01, 2001.

Determining Hit Rate in Pattern Search

Richard J. Bolton, David J. Hand, and Niall M. Adams

Department of Mathematics, Imperial College
180 Queen's Gate, London, UK. SW7 2BZ
{r.bolton, d.j.hand, n.adams}@ic.ac.uk
http://stats.ma.ic.ac.uk/

Abstract. The problem of spurious apparent patterns arising by chance is a fundamental one for pattern detection. Classical approaches, based on adjustments such as the Bonferroni procedure, are arguably not appropriate in a data mining context. Instead, methods based on the false discovery rate - the proportion of flagged patterns which do not represent an underlying reality - may be more relevant. We describe such procedures and illustrate their application on a marketing dataset.

1 Introduction

There are two rather different kinds of data mining activity [1], [2]. One kind, which has a long history and sound theoretical base, is that of the overall descriptive summary or *modelling* of a data set. Tools for this kind of exploration have been rigorously developed by statisticians. They include regression, cluster analysis, and tree methods. The other kind of data mining activity, which has been the subject of formal investigation for only a relatively short period, is *pattern detection*. A *pattern* is a local feature of the data; a property of only a few cases or a small part of the data. Examples of patterns include transient, occasionally repeated, features in EEG trace, outliers, anomaly detection, sudden changes of behaviour, head and shoulders patterns in technical analysis, sequences of nucleotides, bread dominoes in market basket analysis, and so on.

It is perhaps surprising that the theoretical exploration of the concept of pattern is so recent, given that so much of our knowledge is encapsulated in this form. That is, we often try to understand the world via simple patterns. Thus, for example, we know that people who smoke are more likely to suffer from lung cancer, that simultaneous use of a credit card in two different locations suggests fraud, that gravitational microlensing suggests a black hole, and that certain features in meteorology represented by pressure troughs, pressure centres, and so on are predictive of high winds. Rule-based systems (e.g. [3]), which became very popular in the1980s as the architecture underlying expert systems (e.g. [4], [5]) represent one kind of pattern, but patterns are more general than the simple antecedent-consequent form of rules. We speculate that the reason for the relatively recent interest in formal methods for pattern detection is that non-trivial such exercises require substantial computing power, and this has only become available recently. This would also explain why statisticians have

D.J. Hand et al. (Eds.): Pattern Detection and Discovery, LNAI 2447, pp. 36–48, 2002.
© Springer-Verlag Berlin Heidelberg 2002

not developed a theory for this area, and why most of the work has arisen in more computational disciplines, such as data mining, artificial intelligence, and machine learning.

Examination of the pattern detection literature reveals that by far the majority of the publications in the area deal with *methods* for pattern detection - that is, with *algorithms* - and some rather sophisticated and highly efficient tools have been developed for certain application areas. We are thinking, especially, of association analysis and genome analysis. However, relatively little consideration has been given to the statistical properties of patterns. This matters because the extensive pattern search, over a large, often vast, space of possible patterns, means that many of the structures identified as possible patterns will be spurious, in the sense that they will have arisen because of a chance configuration of data points, not representing any underlying reality. We might describe such patterns as 'false' because they are simply chance occurrences. This matters because of the very real danger that many, perhaps most, of the detected patterns will be false. This will lead to mistaken decisions, inappropriate actions, and, of course, a backlash against data mining.

In Section 2 of this paper we describe the background to the problem and introduce our terminology. In Section 3 we describe how the control of false discovery rate may be a more appropriate strategy in data mining than the classic control of familywise error rate. In Section 4 we illustrate with a market basket analysis example. Section 5 illustrates further difficulties which arise when there are large numbers of potential patterns to be examined.

2 Background

In pattern detection data mining, one searches through a huge pattern space, identifying those particular configurations which have a criterion value exceeding some threshold. Criteria such as confidence ([6], [7], [8]) are especially familiar in data mining contexts, though any other measure of departure from a background model can be used (such as log odds, the size of a residual, or an atypicality measure, as in outlier detection). Interest in these patterns rarely lies in the observed data themselves, but is usually aimed at inferences from them. For example, we will generally not be interested in the fact that this year the transactions recorded in our database showed that most people who bought items A and B also bought item C. Rather, we will want to know whether this generalises to other people who might buy items A and B - for example, next year's set of purchasers. Put another way, we will want to know if the criterion value obtained from the available data represents a real underlying relationship, or could have arisen by chance even though the process generating the data includes no such relationship.

This means that the output of our data mining algorithm should include two pieces of information: firstly, the criterion value of the detected A-B-C pattern (e.g. its confidence measure), and, secondly, the probability that the observed criterion value for the pattern could have arisen by chance if there is no real underlying relationship. A sizeable value for the criterion tells that the pattern is worth knowing about, and a small probability tells us that it is unlikely that the pattern is simply an accidental grouping of this year's customers. The second of these notions is formalised in terms of the *significance probability*: the probability that we would observe an A-B-C pattern with a criterion as large as that observed in our data if there was no relationship in the underlying population. If the size of the criterion is above some threshold and the significance probability is below some threshold, then the pattern is deemed worthy of further investigation.

At this point it will help if we introduce some terminology. We define:

- A *pattern* is a vector of data values.
- A *flagged pattern* is one for which the criterion value (e.g. confidence) exceeds the criterion threshold.
- A *significant pattern* is one for which the significance probability is less than a threshold called the significance level. A pattern which is not significant will be termed *non-significant*.
- A *substantive pattern* is one for which there is a genuine propensity for values to be related. A pattern which is not substantive will be termed *non-substantive*.

We will also use the following terms:

- Of the significant patterns, the proportion which are substantive is the *hit rate*.
- Of the significant patterns, the proportion which are not substantive is the *false discovery rate*.

The concept of significance is well known, even if it is rarely taken into account in data mining algorithms. The consequence of ignoring it is that one may identify many patterns as being of potential interest (i.e. flag them) even though it is unlikely that they are substantive (i.e. even though it is unlikely they represent anything of real substance which generalises beyond the particular set of data being analysed). Since such flagged patterns will generally be the subject of detailed investigation, this represents a waste of resources, as well as leading to doubt being cast on the data mining enterprise.

It is also well known that fixing the significance level for individual patterns can lead to identifying many non-substantive patterns as significant. For example, to take an extreme situation, suppose that a population of 100,000 patterns contains none which are substantive, and that each of the 100,000 is tested at the 5% level. Then elementary theory shows that the probability that the algorithm identifies no patterns as significant, which is the ideal result, is 0.95^{10^5}. This is vanishingly small. In general, in these extreme circumstances the probability that k patterns will be detected as significant is $^{10^5}C_k \times 0.05^k \times 0.95^{10^5-k}$. Although this is an extreme example, in general, when there do exist some substantive patterns, controlling for the significance of each pattern individually leads to a situation in which many of the flagged patterns correspond to situations in which there is no underlying structure.

To overcome this problem, rather than fixing the level at which each pattern is tested, multiple comparison procedures have been developed which fix the overall probability that *any* non-substantive pattern will be detected as significant (e.g. [9], [10]). This overall level is termed the *familywise error rate.*

Standard multiple comparison procedures have the opposite disadvantage from fixing the significance level of each individual test. By fixing the overall level, such procedures ensure that the probability of concluding that each individual non-substantive pattern is significant is very low indeed. But it also follows that the probability of detecting a substantive pattern as significant is very small. (Each individual test has low power.) In general, only substantive patterns which have a very large criterion value indeed have more than a tiny chance of being declared significant. Patterns with very large criterion values will generally be already well-known, and the purpose of data mining is not merely to replicate things that are already familiar.

We see from the above that conventional approaches control the probability of declaring at least one non-substantive pattern to be significant. However, in the data mining context, perhaps a more appropriate measure to control would be the proportion of significant patterns which are in fact non-substantive. This tells us what fraction of those patterns we declare as worth investigating further do not reflect a genuine underlying relationship. In economic terms, this may be more relevant: for a given investigation cost, from this we can say what proportion of our significant patterns are likely to be real, and hence what proportion are possibly commercially valuable. We look at these measures and their properties in more detail in the next section.

3 Familywise Error Rate, False Discovery Rate, and Controlling the False Discovery Rate

In Section 2 we suggested that, the proportion of significant patterns which did not represent true underlying structures (i.e. were not substantive) might be more appropriate than the conventional multiple comparisons test criteria for use in data mining. Unfortunately, since the 'underlying reality' is never actually known, we cannot determine this. What we can do, however, following [11] is control the *expected value* of the proportion of significant patterns which represent non-substantive patterns: the *expected false discovery rate.* Adopting the notation shown in Table 1, this is $E\left(b/(b+d)\right)$.

Table 1. Numbers of patterns in each category.

	Non-significant	Significant
Non-substantive	a	b
Substantive	c	d

The familywise error rate of conventional approaches is $P(b > 0)$, the probability that at least one non-substantive hypothesis is rejected. Using these definitions, Benjamini and Hochberg [11] show that $E(b/(b+d)) \leq P(b > 0)$. Building on this result, Benjamini and Hochberg [11] demonstrate that the following simple procedure controls the false discovery rate at q. Suppose there are M patterns, and that the ith has significance probability S_i. Let $O(i)$ be the function which orders these, so that

$$S_{(1)} \leq S_{(2)} \leq .... \leq S_{(M)}$$

where significance probability $S_{(i)}$ corresponds to pattern $O^{-1}(i)$. Then declare as significant all patterns for which $S_{(i)} \leq iq/m$.

4   Market Basket Example

Before we can control the expected false discovery rate, we require probabilities that the patterns discovered have arisen from our baseline distribution. In our example we count the pairwise associations of unique product *types* in 77,425 baskets from a Belgian supermarket. In this data set products of a similar type, e.g. different kinds of apples, were given the same identifier. Identifiers were recorded as distinct numerical codes in order to preserve the anonymity of the product types in the data set for commercial reasons. The baseline we choose here to illustrate the multiplicity phenomenon is that of independence; that is to say, we define a substantive pattern as one that does not occur by chance. In this example, we might expect to see 'uninteresting' patterns as well as interesting ones. The probability assigned to an association reflects its departure from the independence assumption and makes no assumptions about what is interesting to the domain user. In this respect it marks the beginning of the knowledge discovery process, producing a subset of results for the domain expert to examine. DuMouchel [12] looks at drug/event associations, reflecting that the independence assumption is unlikely to hold in this application, and modifying the baseline distribution by introducing a five-parameter prior to a log-linear model. We consider the independence baseline model for pairwise association analysis in this example as it is more likely to hold in market basket analysis; probabilities are easy to interpret and thus help illustrate the multiple comparison phenomenon.

We adopt the notation of [12], but omit the stratification variable. Thus, the number of baskets containing item I_k is denoted by N_k, and the number of baskets containing both items I_k and I_l is denoted by N_{kl}. The baseline frequency (the expected number of baskets containing both items I_k and I_l) is denoted by E_{kl}. Under the independence assumption we estimate this by

$$\hat{E}_{kl} = \frac{N_k N_l}{N} \tag{1}$$

Using arguments from elementary probability, we assume that each basket is independently drawn from a large population of baskets. The distribution of a particular item, I_k, in a basket is Bernoulli with parameter p_k and is the same for each basket in this population, independent of the other items in the basket. If we have N independently distributed baskets, the distribution of N_k, the number baskets that contain item k, is Binomial(N, p_k). If p_k is small and N large then we can approximate N_k by a Poisson(Np_k) or for very large N by Normal(Np_k, $Np_k(1\text{-}p_k)$). We do not know the probabilities p_k, but we can estimate them in the usual way by

$$\hat{p}_k = \sum_{i=1}^{N} \mathbf{I}((x_{I_k})_i = 1)/N$$

or in other words $\hat{p}_k = N_k/N$.

However, we are interested in associations and we would like to compare the number of baskets containing some itemset (of size ≥ 1) with a baseline expectation based on the assumption that the items within this itemset are distributed in baskets independently of each other. Suppose we have an itemset of size two consisting of items I_k and I_l; the distribution of the number of baskets containing both items I_k and I_l, N_{kl}, is Binomial(N, p_{kl}), with possible approximations by Poisson(Np_{kl}) or Normal(Np_{kl}, $Np_{kl}(1\text{-}p_{kl})$). Again, we do not know the underlying probability that a basket contains both items I_k and I_l, p_{kl}, but under the assumption of independence in our baseline distribution we can write $p_{kl} = p_k p_l$. We can thus estimate p_{kl} by $\hat{p}_{kl} = \hat{p}_k \hat{p}_l$ and obtain $\hat{E}_{kl} = N\hat{p}_k\hat{p}_l = N_k N_l/N$ as our estimate of the expected baseline frequency as above.

To obtain probabilities of obtaining more extreme values of N_{kl} than those we have observed, we compare N_{kl} with a Poisson($N_k N_l/N$) distribution.

For comparison purposes we created an artificial data set by taking the marginal counts for each product in the real data set and assigning this number of products randomly over each of 77,425 baskets. For example, if product A occurs in 2,463 baskets in the real data set then it is allocated randomly to the same number of baskets in the artificial data set. Each product is assigned independently of other products in each basket and so we expect no significant associations. The distribution of observed values forms a kind of empirical null distribution with which to compare values from the real data set.

We counted all pairwise associations (N_{kl}) and product margins (N_k) for both the real and artificial basket data sets using the *Apriori* algorithm. There were 2,273,617 different pairwise associations in the artificial data set and 3,244,043 in the real data set (out of a possible 127,560,378). We can see that the real data set contains both very large and very small baskets with a higher frequency than the random data set (Figure 1) and that the null hypothesis of independence may not hold for some associations. This suggests that the simple independence model may not be ideal for this problem, and perhaps one should adopt a model that reflects the observed basket size. Figure 2 shows the distribution of observed and expected numbers of pairwise associations in the real and artificial data sets; the expected numbers of associations were calculated using Equation 1. This figure shows that the observed values of the pairwise associations from the real data set show greater variance than those from the artificial data set representing the independence model. The black crosses (in the region to the top-left of the figure) represent pairwise associations whose observed counts are so high as to be highly unlikely (with probability less than 10^{-16}) to have arisen from the baseline model of independence.

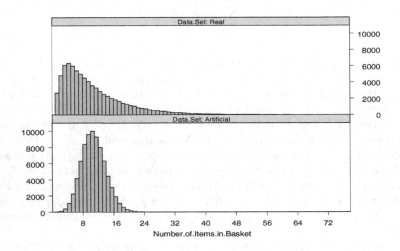

Fig. 1. Histogram of basket sizes for the market basket data.

We applied Benjamini/Hochberg and Bonferroni adjustments to p-values obtained from significance tests from Poisson distributions as described. As there are 127,560,378 possible pairwise associations we must include all of these in the analysis; those associations with zero count need not have p-values calculated explicitly as these will have p-values close to 1: 127,560,378 is simply used as the divisor m in Section 3 above. Table 2 and Table 3 show the number of associations flagged as being significant after the Benjamini/Hochberg adjustment of p-values for the real and artificial data sets respectively. Table 4 and Table 5 show the number of associations flagged as being significant after the Bonferroni adjustment of p-values for the real and artificial data sets respectively.

It is clear that in a data set of this size and with the global model poorly specified (there is evidence to suggest that the independence assumption is incorrect for many thousands of associations, as we alluded to earlier), the Benjamini/Hochberg correction is not conservative enough to reduce to practical levels the number of associations for a domain expert to assess, unless the significance level is made very small indeed. The more conservative Bonferroni adjustment reduces the number of candidate patterns to a more manageable number.

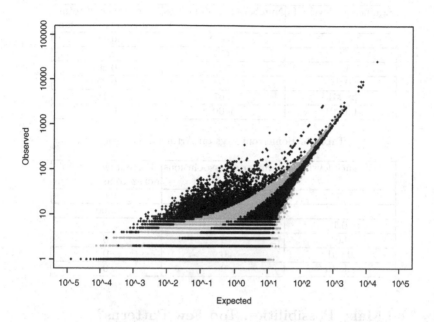

Fig. 2. Plot of observed and expected values for associations. Random basket data (black dots), real basket data (grey dots) with p-values $< 10^{-16}$ (black cross).

Table 2. Benjamini/Hochberg flags on Real data set.

Significance level (%)	Number of associations	Percentage of total non-zero associations
5	810997	25.00
1	346846	10.69
0.1	89050	2.75
0.01	30781	0.95
0.001	15918	0.49
0.0001	11095	0.34

Table 3. Benjamini/Hochberg flags on Artificial data set.

Significance level (%)	Number of associations	Percentage of total non-zero associations
5	246748	10.85
1	65943	2.90
0.1	1884	0.08
0.01	0	0
0.001	0	0
0.0001	0	0

Table 4. Bonferroni flags on Real data set.

Significance level (%)	Number of associations	Percentage of total non-zero associations
5	13459	0.41
1	10780	0.33
0.1	8523	0.26
0.01	7008	0.22
0.001	5867	0.18
0.0001	5000	0.15

Table 5. Bonferroni flags on Artificial data set.

Significance level (%)	Number of associations	Percentage of total non-zero associations
5	118	0.01
1	2	0.00
0.1	0	0
0.01	0	0
0.001	0	0
0.0001	0	0

5   Too Many Possibilities, Too Few Patterns?

Pattern detection data mining aims to detect the classic needle in the haystack. The fact that this can even be contemplated as a real possibility is thanks to the computer. However, the computer is merely a tool, and cannot overcome fundamental structural problems. In this section we illustrate two difficulties which arise from the fact that there are large numbers of potential patterns, but only relative few genuine patterns.

We first use a simple artificial example to demonstrate a fundamental law of diminishing returns which will affect all pattern detection data mining efforts.

In general, whatever strategy one adopts for deciding which patterns to investigate in detail, the general approach will be to compare the criterion value for each putative pattern with a threshold and accept or reject that pattern for further investigation according to whether or not the criterion exceeds the threshold. The lower the threshold, the more patterns which will be chosen for further investigation. Unfortunately, however, there is generally a negative correlation between the number of patterns flagged as worth investigating and the proportion of investigated patterns which turn out to be 'real'.

We shall take the (grossly oversimplified) situation in which there are only two population situations underlying the patterns explored: situations where there are no structures and situations where there are structures, with all these being identical in size. The distribution of values of the pattern criterion will depend on what criterion is used. For example, if the criterion is a proportion of cases showing a certain pattern of values (e.g., a pattern of purchases) then a binomial or Poisson distribution might be appropriate. However, for illustrative purposes, and to keep things simple, we shall suppose that the pattern criterion follows a normal distribution. In particular, we shall suppose that the observed pattern criteria for situations where there is no real structure in the populations takes values which follow a standard normal distribution, $N(0, 1)$, and that the observed pattern criteria for situations where there *is* real structure in the populations takes values which follow a normal distribution with unit variance and mean 1, $N(1, 1)$. For simplicity, we shall call the 'no real structure' cases, 'class 0' cases, and the others 'class 1' cases. Suppose also that n_0 of the patterns explored have no underlying population structure, and n_1 of them do have an underlying population structure. We can let $n_0 = kn_1$, where $k > 0$ is a value indicating the ratio of the two population samples; setting $k < 1$ implies that there are more patterns with real structure whereas setting $k > 1$ implies *fewer* patterns with real structure – the more likely scenario.

Detected apparent patterns are flagged as being worthy of closer investigation if their criterion value exceeds some threshold. For any given threshold value, let $n_{1|1}$ represent the number of class 1 (real structure) cases which have criterion value exceeding the threshold, and $n_{1|0}$ represent the number of class 0 (no real structure) cases which have criterion value exceeding the threshold. Then the proportion of real structure cases amongst those exceeding the threshold is $n_{1|1}/(n_{1|0} + n_{1|1})$. Figure 3 shows this proportion plotted against a threshold value for various values of k.

We see that, as the threshold value increases, so the proportion belonging to the 'real structure' class increases. However, as the ratio of real to non-real structure decreases (k increases), we see that the proportion belonging to the real structure class, above a particular threshold value, decreases.

In practical terms this means is that it is better to restrict oneself to recommending just the few largest patterns. The more one relaxes this, and the more patterns one recommends investigating in detail, so the lower one's hit rate will become, especially when there are relatively few real patterns in the data: there is a pronounced law of diminishing returns.

There is a secondary cautionary note that might be appended here: it seems probable that, the greater the difference between the threshold and the criterion value, the more likely it is that the pattern will already be known. We have already seen some evidence that pattern detection is suffering from this phenomenon. For example, the observation, reported in a real study, that the number of married men is almost exactly equal to the number of married women, and the discovery, again in a real study, that almost all money transfers to US banks occur in dollars, do little to inspire one's confidence in pattern detection as a useful tool.

The example in this section relates especially to measures of the effectiveness of pattern detection algorithms. In the above, we have described the false discovery rate as an appropriate tool, but others report other measures. In particular, referring to Table 1, popular measures of performance are the ratios $a/(a + b)$ and $b/(c + d)$. (In epidemiology, in particular, these are very popular, and corre-

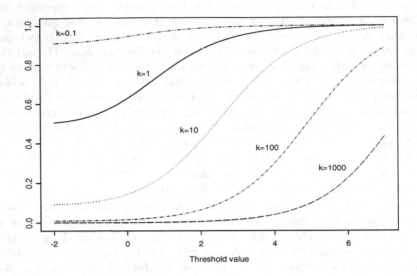

Fig. 3. The proportion of real structure cases amongst those exceeding the threshold value for different values of k.

spond to estimates of the sensitivity and specificity of a screening or diagnostic tool, respectively ([13])). That these can be grossly misleading is seen from the numerical example given in Table 6. Using these numbers, we see that:

1) 20000/20500 = 97.6% of the non-substantive patterns are not flagged as worth investigating.
2) 90/99 = 90.9% of the substantive patterns are flagged as worth investigating. Both of these results are very good, and perhaps suggest that the algorithm is highly effective. However,
3) 500/590 = 84.7% of those flagged as worth investigating are in fact non-substantive (this is the false discovery rate).

The superficial contradiction between the popular measures of sensitivity and specificity and the relevant measure of false discovery rate arises because of the very low prevalence (0.05%) of the putative patterns which are substantive.

Table 6. Numbers of patterns in each category.

	Not flagged	Flagged
Non-substantive	20000	500
Substantive	10	90

6   Conclusions

Pattern detection data mining must face the problem of multiplicity: by chance alone, amongst the vast numbers of potential patterns examined, we would expect to find some, perhaps many, which are anomalously high. This is in addition to spurious patterns arising from distorted data. Without tools for handling this problem, a large number, perhaps even the majority of detected apparent patterns, will be chance phenomena of the data, and will not reflect any true underlying reality. There are several dangers associated with this. At a low level, it will mean that detected apparent patterns which lead to expenditure and investment, subsequently turn out to be false. Moreover, there will be additional cultural and societal consequences. Recent examples are the fear of a link between MMR vaccinations and autism, and racial discrimination in efforts to detect potential terrorists boarding aircraft. Other examples are not hard to find.

Classical approaches to this problem are based on multiple comparison procedures, such as the Bonferroni adjustment for familywise error rate. However, this procedure can be overwhelmed by the large numbers of putative patterns examined in data mining. In this paper we have examined an alternative approach, based on the (more relevant) false discovery rate, rather than the familywise error rate.

Acknowledgements. The work of Richard J. Bolton and David J. Hand described in this paper was supported by EPSRC ROPA grant GR/N08704. We would like to thank Tom Brijs for providing the Belgian supermarket data set used in this paper.

References

1. Hand D.J.: Data mining: statistics and more? The American Statistician **52** (1998) 112–118
2. Hand D.J., Blunt G., Kelly M.G., Adams N.M.: Data mining for fun and profit. Statistical Science **15** (2000) 111–131
3. Waterman D.A., Hayes-Roth F.: Pattern-Directed Inference Systems. Academic Press, New York (1978)
4. Buchanan B.G., Shortliffe E.H. (eds.): Rule-Based Expert Systems. Addison-Wesley, Reading, Massachusetts (1984)
5. Klahr P., Waterman D.A.: Expert Systems: Techniques, Tools, and Applications. Addison-Wesley, Reading, Massachusetts (1986)
6. Agrawal R., Imielinski T., Swami A.: Mining association rules between sets of items in large databases. In: Proceedings of the ACM SIGMOD Conference, Washington DC. (1993)
7. Agrawal R., Imielinski T., Swami A.: Database mining: a performance perspective. IEEE Transactions on Knowledge and Data Engineering **5** (1993) 914–925
8. Hand D.J., Blunt G., Bolton R.J.: A note on confidence and support. Technical Report, Department of Mathematics, Imperial College, London (2001)

9. Miller R.G.: Simultaneous Statistical Inference. 2nd ed. Springer-Verlag, New York (1981)
10. Pigeot I.: Basic concepts of multiple tests – a survey. Statistical Papers **41** (2000) 3–36
11. Benjamini Y., Hochberg Y.: Controlling the false discovery rate. Journal of the Royal Statistical Society, Series B **57** (1995) 289–300
12. DuMouchel, W.: Bayesian data mining in large frequency tables, with an application to the FDA Spontaneous Reporting System. The American Statistician **53** (1999) 177–202
13. Hand D.J.: Construction and Assessment of Classification Rules. Wiley, Chichester (1997)

An Unsupervised Algorithm for Segmenting Categorical Timeseries into Episodes

Paul Cohen[1], Brent Heeringa[1], and Niall M. Adams[2]

[1] Department of Computer Science. University of Massachusetts, Amherst. Amherst, MA 01003
{cohen | heeringa}@cs.umass.edu
[2] Department of Mathematics. Imperial College. London, UK
n.adams@ic.ac.uk

Abstract. This paper describes an unsupervised algorithm for segmenting categorical time series into episodes. The VOTING-EXPERTS algorithm first collects statistics about the frequency and boundary entropy of ngrams, then passes a window over the series and has two "expert methods" decide where in the window boundaries should be drawn. The algorithm successfully segments text into words in four languages. The algorithm also segments time series of robot sensor data into subsequences that represent episodes in the life of the robot. We claim that VOTING-EXPERTS finds meaningful episodes in categorical time series because it exploits two statistical characteristics of meaningful episodes.

1 Introduction

Though we live in a continuous world, we have the impression that experience comprises episodes: writing a paragraph, having lunch, going for a walk, and so on. Episodes have hierarchical structure; for instance, writing a paragraph involves thinking of what to say, saying it, editing it; and these are themselves episodes. Do these examples of episodes have anything in common? Is there a domain-independent, formal notion of episode sufficient, say, for an agent to segment continuous experience into meaningful units?

One can distinguish three ways to identify episode boundaries: First, they may be *marked*, as spaces mark word boundaries and promoters mark coding regions in DNA. Second, episodes may be *recognized*. For instance, we recognize nine words in the sequence "itwasabrightcolddayinapriland". Third we might *infer* episode boundaries given the statistical structure of a series. For example, "juxbtbcsjhiudpmeebzjobqsjmboe" is formally (statistically) identical with "itwasabrightcolddayinapriland" — one is obtained from the other by replacing each letter with the adjacent one in the alphabet — however, the latter is easily segmented by recognition whereas the former requires inference.

This paper proposes two statistical characteristics of episode boundaries and reports experiments with an unsupervised algorithm called VOTING-EXPERTS based on these characteristics. We offer the conjecture that these characteristics are domain-independent and illustrate the point by segmenting text in four languages.

D.J. Hand et al. (Eds.): Pattern Detection and Discovery, LNAI 2447, pp. 49–62, 2002.
© Springer-Verlag Berlin Heidelberg 2002

2   The Episode Boundary Problem

Suppose we remove all the spaces and punctuation from a text, can an algorithm figure out where the word boundaries should go? Here is the result of running VOTING-EXPERTS on the first 500 characters of George Orwell's *1984*. The $\star$ symbols are induced boundaries:

> Itwas $\star$ a $\star$ bright $\star$ cold $\star$ day $\star$ in $\star$ April $\star$ andthe $\star$ clockswere $\star$ st $\star$ ri $\star$
> king $\star$ thi $\star$ rteen $\star$ Winston $\star$ Smith $\star$ his $\star$ chin $\star$ nuzzl $\star$ edinto $\star$ his $\star$ brea
> $\star$ st $\star$ in $\star$ aneffort $\star$ to $\star$ escape $\star$ the $\star$ vilewind $\star$ slipped $\star$ quickly $\star$ through
> $\star$ the $\star$ glass $\star$ door $\star$ sof $\star$ Victory $\star$ Mansions $\star$ though $\star$ not $\star$ quickly $\star$
> en $\star$ ought $\star$ oprevent $\star$ aswirl $\star$ ofgrit $\star$ tydust $\star$ from $\star$ ent $\star$ er $\star$ inga $\star$
> long $\star$ with $\star$ himThe $\star$ hall $\star$ ways $\star$ meltof $\star$ boiled $\star$ cabbage $\star$ and $\star$ old $\star$
> ragmatsA $\star$ tone $\star$ endof $\star$ it $\star$ acoloured $\star$ poster $\star$ too $\star$ large $\star$ for $\star$ indoor
> $\star$ dis $\star$ play $\star$ hadbeen $\star$ tack $\star$ ed $\star$ tothe $\star$ wall $\star$ It $\star$ depicted $\star$ simplya $\star$
> n $\star$ enormous $\star$ face $\star$ more $\star$ than $\star$ ametre $\star$ widethe $\star$ faceof $\star$ aman $\star$ of $\star$
> about $\star$ fortyfive $\star$ witha $\star$ heavy $\star$ black $\star$ moustache $\star$ and $\star$ rugged $\star$ ly $\star$
> handsome $\star$ featur

The segmentation is imperfect: Words are run together (*Itwas, aneffort*) and broken apart (*st $\star$ ri $\star$ king*). Occasionally, words are split between segments (*to* in *en $\star$ ought $\star$ oprevent*). Still, the segmentation is surprisingly good when one considers that it is based on nothing more than statistical features of subsequences of letters — not words, as no word boundaries are available — in Orwell's text.

How can an algorithm identify subsequences that are *meaningful* in a domain lacking any knowledge about the domain; and particularly, lacking positive and negative training instances of meaningful subsequences? VOTING-EXPERTS must somehow detect *domain-independent* indicators of the boundaries of meaningful subsequences. In fact, this is a good description of what it does. It implements a weak theory of domain-independent features of meaningful units. The first of these features is that entropy remains low inside meaningful units and increases at their boundaries; the second is that high-frequency subsequences are more apt to be meaningful than low-frequency ones.

3   Characteristics of Episodes

The features of episodes that we have implemented in the VOTING-EXPERTS algorithm are called *boundary entropy* and *frequency*:

Boundary entropy. Every unique subsequence is characterized by the distribution of subsequences that follow it; for example, the subsequence "en" in this sentence repeats seven times and is followed by tokens c (4 times), t, s and ", a distribution of symbols with an entropy value (1.66, as it happens). In general, every subsequence S has a boundary entropy, which is the entropy of the distribution of subsequences of length m that follow it. If S is an episode, then the boundary entropies of subsequences of S will have an interesting profile: They will start relatively high, then sometimes drop, then peak at the last element of S. The reasons for this are first, that the predictability of elements within an episode increases as the episode extends over time; and second, the elements that immediately follow an episode are relatively uncertain. Said differently, within episodes, we know roughly what will happen, but at episode boundaries we become uncertain.

Frequency. Episodes, recall, are meaningful sequences. They are patterns in a domain that we call out as special, important, valuable, worth committing to memory, worth naming, etc. One reason to consider a pattern meaningful is that one can use it for something, like prediction. (Predictiveness is another characteristic of episodes nicely summarized by entropy.) Rare patterns are less useful than common ones simply because they arise infrequently, so all human and animal learning places a premium on frequency. In general, episodes are common patterns, but not all common patterns are episodes.

4 Related Work

Many methods have been developed for segmenting time series. Of these, many deal with continuous time series, and are not directly applicable to the problem we are considering here. Some methods for categorical series are based on compression (e.g., [1]), but compression alone finds common, not necessarily meaningful, subsequences. Some methods are trained to find instances of patterns or templates (e.g., [2,3]) or use a supervised form of compression (e.g., [4]), but we wanted an unsupervised method. There is some work on segmentation in the natural language and information retrieval literature, for instance, techniques for segmenting Chinese, which has no word boundaries in its orthography, but again, these methods are often supervised. The method in [5] is similar to ours, though it requires supervised training on very large corpora. The parsing based on mutual information statistics approach in [6] is similar to our notion of boundary entropy. [7] provides a developmentally plausible unsupervised algorithm for word segmentation, but the procedure assumes known utterance boundaries. [8] give an unsupervised segmentation procedure for Japanese, however it too supposes known sequence boundaries. With minor alterations, their segmentation technique is applicable to our domain, but we found that VOTING-EXPERTS consistently outperforms it. We know of no related research on characteristics of meaningful episodes, that is, statistical markers of boundaries of meaning-carrying subsequences.

5 The Voting Experts Algorithm

VOTING-EXPERTS includes experts that attend to boundary entropy and frequency and is easily extensible to include experts that attend to other characteristics of episodes. The algorithm simply moves a window across a time series and asks for each location in the window whether to "cut" the series at that location. Each expert casts a vote. Each location takes n steps to traverse a window of size n, and is seen by the experts in n different contexts, and may accrue up to n votes from each expert. Given the results of voting, it is a simple matter to cut the series at locations with high vote counts. Here are the steps of the algorithm:

Build an ngram trie of depth $n + 1$. Nodes at level $i + 1$ of the trie represent ngrams of length i. The children of a node are the extensions of the ngram represented by the node. For example, $a\ b\ c\ a\ b\ d$ produces the following trie of depth 3:

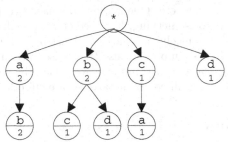

Every ngram of length 2 or less in the sequence $a\ b\ c\ a\ b\ d$ is represented by a node in this tree. The numbers in the lower half of the nodes represent the frequencies of the subsequences. For example, the subsequence ab occurs twice, and every occurrence of a is followed by b.

For the first 10,000 characters in Orwell's text, an ngram trie of depth 8 includes 33774 nodes, of which 9109 are leaf nodes. That is, there are over nine thousand unique subsequences of length 7 in this sample of text, although the average frequency of these subsequences is 1.1—most occur exactly once. The average frequencies of subsequences of length 1 to 7 are 384.4, 23.1, 3.9, 1.8, 1.3, 1.2, and 1.1.

Calculate boundary entropy. The boundary entropy of an ngram is the entropy of the distribution of tokens that can extend the ngram. The entropy of a distribution for a discrete random variable X is

$$-\sum_{x \in X} p(x) \log p(x)$$

Boundary entropy is easily calculated from the trie. For example, the node a in the tree above has entropy equal to zero because it has only one child, ab, whereas the entropy of node b is 1.0 because it has two equiprobable children, bc and bd. Clearly, only the first n levels of the ngram tree of depth $n + 1$ can have node entropy scores.

Standardize frequencies and boundary entropies. In most domains, there is a systematic relationship between the length and frequency of patterns; in general, short patterns are more common than long ones (e.g., on average, for subsets of 10,000 characters from Orwell's text, 64 of the 100 most frequent patterns are of length 2; 23 are of length 3, and so on). Our algorithm will compare the frequencies and boundary entropies of ngrams of different lengths, but in all cases we will be comparing how *unusual* these frequencies and entropies are, relative to other ngrams of the same length. To illustrate, consider the words "a" and "an." In the first 10000 characters of Orwell's text, "a" occurs 743 times, "an" 124 times, but "a" occurs only a little more frequently than other one-letter ngrams, whereas "an" occurs much more often than other two-letter ngrams. In this sense, "a" is ordinary, "an" is unusual. Although "a" is much more common than "an" it is much less unusual relative to other ngrams of the same length. To capture this notion, we standardize the frequencies and boundary entropies of the ngrams. To standardize a value in a sample, subtract the sample mean from the value and divide by the sample standard deviation. This has the effect of expressing the value as the number of standard deviations it is away from the sample mean. Standardized, the frequency of "a" is 1.1, whereas the frequency of "an" is 20.4. In other words, the frequency of "an" is 20.4 standard deviations above the mean frequency for sequences of the same length. We standardize boundary entropies in the same way, and for the same reason.

Score potential segment boundaries. In a sequence of length k there are $k - 1$ places to draw boundaries between segments, and, thus, there are 2^{k-1} ways to divide the sequence into segments. Our algorithm is greedy in the sense that it considers just $k - 1$, not 2^{k-1}, ways to divide the sequence. It considers each possible boundary in order, starting at the beginning of the sequence. The algorithm passes a window of length n over the sequence, halting at each possible boundary. All of the locations within the window are considered, and each garners zero or one vote from each expert. Because we have two experts, for boundary-entropy and frequency, respectively, each possible boundary may accrue a maximum of $2n$ votes. This is illustrated below.

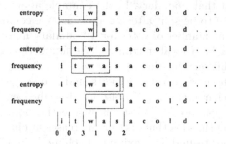

A window of length 3 is passed along the sequence itwasacold. Initially, the window covers itw. The entropy and frequency experts each decide where they could best insert a boundary within the window (more on this, below). The entropy expert favors the boundary between t and w, while the frequency expert favors the boundary between w and whatever comes next. Then the window moves one location to the right and the process repeats. This time, both experts decide to place the boundary between t and w. The window moves again and both experts decide to place the boundary after s, the last token in the window. Note that each potential boundary location (e.g., between t and w) is seen n times for a window of size n, but it is considered in a slightly different context each time the window moves. The first time the experts consider the boundary between w and a, they are looking at the window itw, and the last time, they are looking at `was`. In this way, each boundary gets up to $2n$ votes, or $n = 3$ votes from each of two experts. The `wa` boundary gets one vote, the `tw` boundary, three votes, and the `sa` boundary, two votes.

The experts use slightly different methods to evaluate boundaries and assign votes. Consider the window `itw` from the viewpoint of the boundary entropy expert. Each location in the window bounds an ngram to the left of the location; the ngrams are `i`, `it`, and `itw`, respectively. Each ngram has a standardized boundary entropy. The boundary entropy expert votes for the location that produces the ngram with the highest standardized boundary entropy. As it happens, for the ngram tree produced from Orwell's text, the standardized boundary entropies for `i`, `it`, and `itw` are 0.2, 1.39 and 0.02, so the boundary entropy expert opts to put a boundary after the ngram `it`.

The frequency expert places a boundary so as to maximize the sum of the standardized frequencies of the ngrams to the left and the right of the boundary. Consider the window `itw` again. If the boundary is placed after `i`, then (for Orwell's text) the standardized frequencies of `i` and `tw` sum to 1.73; if the boundary is placed after `it`, then the standardized frequencies of `it` and `w` sum to 2.9; finally, if it is placed after `itw`, the algorithm has only the standardized frequency of `itw` to work with; it is 4.0. Thus, the frequency expert opts to put a boundary after `itw`.

Segment the sequence. Each potential boundary in a sequence accrues votes, as described above, and now we must evaluate the boundaries in terms of the votes and decide where to segment the sequence. Our method is a familiar "zero crossing" rule: If a potential boundary has a locally maximum number of votes, split the sequence at that boundary. In the example above, this rule causes the sequence `itwasacold` to be split after `it` and `was`. We confess to one embellishment on the rule: The number of votes for a boundary must exceed an absolute threshold, as well as be a local maximum. We found that the algorithm splits too often without this qualification.

Let us review the design of the experts and the segmentation rule, to see how they test the characteristics of episodes described earlier. The boundary entropy expert assigns votes to locations where the boundary entropy peaks locally, implementing the idea that entropy increases at episode boundaries. The frequency expert tries to find a "maximum likelihood tiling" of the sequence, a placement of boundaries that makes the ngrams to the left and right of the boundary as likely as possible. When both experts vote for a boundary, and especially when they vote repeatedly for the same boundary, it is likely to get a locally-maximum number of votes, and the algorithm is apt to split the sequence at that location.

6   Evaluation

In these experiments, induced boundaries stand in six relationships to episodes.

1. The boundaries coincide with the beginning and end of the episode;

2. The episode falls entirely within the boundaries and begins or ends at one boundary.
3. The episode falls entirely within the boundaries but neither the beginning nor the end of the episode correspond to a boundary.
4. One or more boundaries splits an episode, but the beginning and end of the episode coincide with boundaries.
5. Like case 4, in that boundaries split an episode, but only one end of the episode coincides with a boundary.
6. The episode is split by one or more boundaries and neither end of the episode coincides with a boundary.

These relationships are illustrated graphically in Figure 1, following the convention that horizontal lines denote actual episodes, and vertical lines denote induced boundaries. The cases can be divided into three groups. In cases 1 and 4, boundaries correspond to both ends of the episode; in cases 2 and 5, they correspond to one end of the episode; and in cases 3 and 6, they correspond to neither end. We call these cases *exact*, *dangling*, and *lost* to evoke the idea of episodes located exactly, dangling from a single boundary, or lost in the region between boundaries.

We use both hit and false-positive rates to measure the accuracy of our episode finding algorithms. To better explain the trade-offs between hits and false-positives we employ the F-measure [9]. This standard comparison metric finds the harmonic mean between precision and recall is defined as

$$\text{F-measure} = \frac{2 \times \text{Precision} \times \text{Recall}}{\text{Precision} + \text{Recall}}$$

where Recall is the hit-rate and Precision is the ratio of correct hits to proposed hits. Note that the difference in proposed and correct hits yields the number of false positives. Higher F-measures indicate better overall performance.

For control purposes we compare VOTING-EXPERTS with two naive algorithms. The first generates a random, sorted sequence of boundaries that is

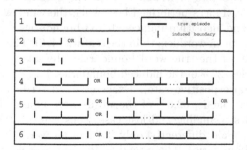

Fig. 1. A graphical depiction of the relationships between boundaries and episodes. Horizontal lines denote true episodes; their ends the correct boundaries. Vertical lines denote induced episode boundaries.

equal in size to the actual number of episodes. We call this algorithm RANDOM-SAMPLE. The second algorithm induces a boundary at every location. We call this algorithm ALL-LOCATIONS.

In many of these experiments, we compare the results of VOTING-EXPERTS with another unsupervised algorithm, SEQUITUR, which also finds structure in categorical time series. SEQUITUR is a compression-based algorithm that builds a context-free grammar from a string of discrete tokens [1]. It has successfully identified structure in both text and music. This structure is denoted by the rules of the induced grammar. Expanding the rules reveals boundary information. In our experiments, expanding only the rule associated with the start symbol – what we refer to as level 1 expansion – most often gives the highest F-measure.

6.1   Word Boundaries

We removed spaces and punctuation from texts in four languages and assessed how well VOTING-EXPERTS could induce word boundaries. We take word boundaries as our gold standard for meaning-carrying units in text because they provide, in most cases, the most unambiguous and uncontentious denotation of episodes. Clearly word prefixes and suffixes might also carrying meaning, but most humans would likely segment a discrete stream of text into words.

Table 1. Results of running four different algorithms on George Orwell's *1984*.

Algorithm	F-measure	Hit Rate	F.P. Rate	Exact %	Dangling %	Lost %
VOTING-EXPERTS	.76	.80	.27	.63	.34	.03
SEQUITUR	.58	.58	.43	.30	.56	.14
ALL-LOCATIONS	.36	1.0	.78	1.0	0.0	0.0
RANDOM-SAMPLE	.21	.22	.79	.05	.34	.61

English. We ran VOTING-EXPERTS, SEQUITUR, and both naive algorithms on the first 50,000 characters of Orwell's *1984*. The detailed results are given in Table 1. VOTING-EXPERTS performed best when the window length was 7 and the threshold 4. The algorithm induced 12153 boundaries, for a mean episode length of 4.11. The mean word length in the text was 4.49. The algorithm induced boundaries at 80% of the true word boundaries (the hit rate) missing 20% of the word boundaries. 27% of the induced boundaries did not correspond to word boundaries (the false positive rate). Exact cases, described above, constitute 62.6% of all cases; that is, 62.6% of the words were bounded at both ends by induced boundaries. Dangling and lost cases constitute 33.9% and 3.3% of all cases, respectively. Said differently, only 3.3% of all words in the text got lost between episode boundaries. These tend to be short words, in fact, 59% of the lost words have length 3 or shorter and 85% have length 5 or shorter. In contrast, all 89% of the words for which the algorithm found exact boundaries are of length 3 or longer.

SEQUITUR performed best when expanding only to the level 1 boundaries. That is, it achieved its highest F-measure by not further expanding any non-terminals off the sentential production. Expanding to further levels leads to a substantial increase in the false positive rate and hence the overall decrease in F-measure. For example, when expanding to level 5, SEQUITUR identified 78% of the word boundaries correctly, 20% dangling and only 2% missed. This happens because it is inducing more boundaries. In fact, at level 5, the false-positive rate of 68% is near the 78% maximum false positive rate achieved by ALL-LOCATIONS. The same behavior occurs to a smaller extent in VOTING-EXPERTS when the splitting threshold is decreased. For example, with a window length of 4 and a threshold of 2, VOTING-EXPERTS finds 74% of the word boundaries exactly but the F-measure decreases because a corresponding increase in the false-positive rate. In general, SEQUITUR found likely patterns, but these patterns did not always correspond to word boundaries.

It is easy to ensure that all word boundaries are found, and no word is lost: use ALL-LOCATIONS to induce a boundary between each letter. However, this strategy induces a mean episode length of 1.0, much shorter than the mean word length. The false-positive count equals the total number of non-boundaries in the text and the false-positive rate converges to the ratio of non-boundaries to total locations (.78). In contrast, VOTING-EXPERTS finds roughly the same number of episodes as there are words in the text and loses very few words between boundaries. This success is evident in the high F-measure (.76) achieved by VOTING-EXPERTS. Not surprisingly, RANDOM-SAMPLE performed poorest on the text.

The appropriate control conditions for this experiment were run and yielded the expected results: VOTING-EXPERTS performs marginally less well when it is required to segment text it has not seen. For example, if the first 10,000 characters of Orwell's text are used to build the ngram tree, and then the algorithm is required to segment the next 10,000 characters, there is a very slight decrement in performance. The algorithm performs very poorly given texts of random words, that is, subsequences of random letters. The effects of the corpus size and the window length are shown in the following graph. The proportion of "lost" words (cases 3 and 6, above) is plotted on the vertical axis, and the corpus length is plotted on the horizontal axis. Each curve in the graph corresponds to a window length, k. The proportion of lost words becomes roughly constant for corpora of length 10,000 and higher.

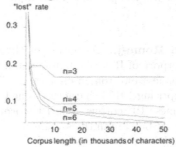

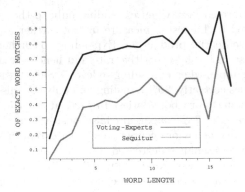

Fig. 2. A comparison of exact match-rate on a per-word basis between SEQUITUR and VOTING-EXPERTS.

Said differently, corpora of this length seem to be required for the algorithm to estimate boundary entropies and frequencies accurately. As to window length, recall that a window of length n means each potential boundary is considered n times by each expert, in n different contexts. Clearly, it helps to increase the window size, but the benefit diminishes.

Further evidence of VOTING-EXPERTS ability to find meaningful word boundaries is given in Figures 2 and 3. In Figure 2 we graph the percentage of exact word matches as a function of word length. For example, SEQUITUR exactly matches 30% of words having length 15 while VOTING-EXPERTS matches 70%. The curves converge at word length 17 because only two words in our corpus have length 17 and both algorithms find only one of them. The curves roughly mimic each other except in the word length interval from 2 to 4. In this period, VOTING-EXPERTS accelerates over SEQUITUR because it finds disproportionately more exact matches than SEQUITUR. This phenomenon is even easier to see in Figure 3. Here cumulative percentage of exact word matches is plotted as a function of word lengths and the distribution of word lengths is given behind the curves. The slope of VOTING-EXPERTS is steeper than SEQUITUR in the interval from 2 to 4 revealing the success it has on the most frequent word lengths. Furthermore, words with length 2, 3, and 4 comprise over 57% of the Orwell corpus, so at places where accuracy is perhaps most important, VOTING-EXPERTS performs well.

Chinese, German and Roma-ji. As a test of the generality of VOTING-EXPERTS, we ran it on corpora of Roma-ji, Chinese and German texts. Roma-ji is a transliteration of Japanese into roman characters. The Roma-ji corpus was a set of Anime lyrics comprising 19163 characters. The Chinese text comes from Guo Jim's Mandarin Chinese PH corpus. The PH corpus is taken from stories in newspaper texts and is encoded in in the standard GB-scheme. Franz Kafka's *The Castle* in the original German comprised the final text. For comparison

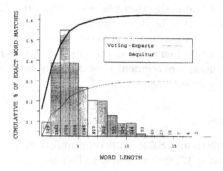

Fig. 3. A comparison of cumulative exact match-rate over word length for SEQUITUR and VOTING-EXPERTS. The background histogram depicts the distribution of word lengths in the Orwell corpus.

purposes we selected the first 19163 characters of Kafka's text and the same number of characters from *1984* and the PH corpus. As always, we stripped away spaces and punctuation, and the algorithm induced word boundaries. The window length was 6. The results are given in Table 2.

Table 2. Results of running VOTING-EXPERTS on Franz Kafka's *The Castle*, Orwell's 1984, a subset of the Chinese PH corpus of newspaper stories, and a set of Roma-ji Anime lyrics.

VOTING-EXPERTS	F-measure	Hit Rate	F.P. Rate	Exact %	Dangling %	Lost %
German	.75	.79	.31	.61	.25	.04
English	.71	.76	.33	.58	.38	.04
Roma-ji	.65	.64	.34	.37	.53	.10
Chinese	.57	.42	.07	.13	.57	.30

Clearly the algorithm is not biased to do well on English. In particular, it performs very well on Kafka's text, losing only 4% of the words and identifying 61% exactly. The algorithm performs less well with the Roma-ji text; it identifies fewer boundaries accurately (i.e., places 34% of its boundaries within words) and identifies fewer words exactly. VOTING-EXPERTS performed worst on Chinese corpus. Only 42% of the boundaries were identified although the false positive rate is an extremely low 7%. The explanation for these results has to do with the lengths of words in the corpora. We know that the algorithm loses disproportionately many short words. Words of length 2 make up 39% of the Chinese corpus, 32% of the Roma-ji corpus, 17% of the Orwell corpus, and 10% of the Kafka corpus, so it is not surprising that the algorithm performs worst on the Chinese corpus and best on the Kafka corpus.

If we incorporate the knowledge that Chinese words are rather short in length by decreasing the splitting threshold, we can increase the F-measure of VOTING-EXPERTS to 77% on the PH corpus. In general, knowledge of the mean episode length can help improve the boundary detection of VOTING-EXPERTS. Like [8], pretraining on a small amount of segmented text may be sufficient to find suitable window and threshold values.

6.2   Robot Episodes

We ran VOTING-EXPERTS and SEQUITUR on a multivariate timeseries of robot controller data comprising 17788 time steps and 65 unique states. Each state was mapped to a unique identifier, and these tokens were given to the algorithm as input. The timeseries data was collected with a Pioneer 2 mobile robot, equipped with sonar and a Sony pan-tilt-zoom camera. The robot wandered around a room-size playpen for 30 minutes looking for interesting objects. Upon finding an object, the robot orbited it for a few minutes. The multivariate timeseries consisted of eight binary variables representing different controllers in our agent architecture. Each variable is 1 when its corresponding controller is active and 0 when its inactive, so potentially, we have $2^8 = 256$ different states, but as mentioned earlier, only 65 manifested during the experiment.

- MOVE-FORWARD
- TURN
- COLLISION-AVOIDANCE
- VIEW-INTERESTING-OBJECT
- RELOCATE-INTERSTING-OBJECT
- SEEK-INTERESTING-OBJECT
- CENTER-CHASIS-ON-OBJECT
- CENTER-CAMERA-ON-OBJECT

This timeseries can be broken up into five different observable robot behaviors. Each behavior represents a qualitatively different episode in the timeseries. We denote these episodes as

- FLEEING
- WANDERING
- AVOIDING
- ORBITING-OBJECT
- APPROACHING-OBJECT

Table 3 summarizes the results of running VOTING-EXPERTS and SEQUITUR on the robot controller data. The definition of hit-rate and false-positive rate is slightly different here. Because the controller data can be noisy at the episode boundaries, we allow *hits* a window of length 1 in either temporal direction. For example, if we induce a boundary at location 10, but the actual boundary is at location 9, we still count it as a hit. We also enforce a rule that actual boundaries can only count once toward induced boundaries. For example, if we induce a boundary at 8 and count it as a hit toward the actual boundary 9, the induced boundary at 10 can no longer count toward 9.

The mean episode length in the robot controller data is 7.13. This length is somewhat smaller than expected because the robot often gets caught up in the corners of its playpen for periods of time and performs a series of wandering, avoiding, and fleeing behaviors to escape. The total number of true episodes was 2491. VOTING-EXPERTS induced 3038 episodes with a hit rate of 66% and a false-positive rate of 46% for a combined F-measure of 59%. Like on Orwell, VOTING-EXPERTS consistently outperforms SEQUITUR on the F-measure. SE-QUITUR does best when expanding to the level 1 boundaries. The transition from level 1 to level 2 produces a sharp increase in the false-positive rate with a corresponding increase in hit rate, however the F-measure decreases slightly. At level 5, SEQUITUR loses only 8% of the episodes but its false-positive rate is 78%, which is near the maximum possible rate of 86%.

Table 3. Results of running SEQUITUR and VOTING-EXPERTS on 30 minutes of robot controller data.

Robot Data	F-measure	Hit Rate	F.P. Rate	Exact %	Dangling Rate	Lost Rate
SEQUITUR						
Level 1	.55	.57	.47	.17	.37	.46
Level 2	.51	.77	.62	.34	.37	.29
Level 3	.32	.88	.71	.48	.33	.19
Level 4	.38	.94	.76	.56	.32	.12
Level 5	.36	.97	.78	.63	.29	.08
VOTING-EXPERTS						
Depth 7, Threshold 4	.59	.66	.46	.20	.39	.41
Depth 9, Threshold 6	.59	.60	.41	.18	.38	.44
Depth 5, Threshold 2	.56	.80	.56	.27	.42	.31

7   Conclusion

For an agent to generalize its experiences, it must divide them into meaningful units. The VOTING-EXPERTS algorithm uses statistical properties of categorical time series to segment them into episodes without supervision or prior training. Although the algorithm does not use explicit knowledge of words or robot behaviors, it detects episodes in these domains. The algorithm successfully segments texts into words in four languages. With less success, VOTING-EXPERTS segments robot controller data into activities. In the future we will examine how other, domain-independent experts can help improve performance. Additionally we are interested in unifying the frequency and boundary entropy experts to more accurately capture the balance of strengths and weaknesses of each method. On a related note, we could employ supervised learning techniques to learn a weigh parameter for the experts, however we favor the unification approach because it removes a parameter from the algorithm and keeps the method completely un-supervised, The idea that meaningful subsequences differ from meaningless ones in some formal characteristics—that syntactic criteria might help us identify semantic units—has practical as well as philosophical implications.

Acknowledgments. We are grateful to Ms. Sara Nishi for collecting the corpus of Anime lyrics. This research is supported by DARPA under contract numbers DARPA/USASMDCDASG60-99-C-0074 and DARPA/AFRLF30602-01-2-0580. The U.S. Government is authorized to reproduce and distribute reprints for gov-ernmental purposes notwithstanding any copyright notation hereon. The views and conclusions contained herein are those of the authors and should not be in-terpreted as necessarily representing the official policies or endorsements either expressed or implied, of DARPA or the U.S. Government.

References

1. Nevill-Manning, C.G., Witten, I. H.: Identifying hierarchical structure in sequences: A linear-time algorithm. Journal of Artificial Intelligence Research **7** (1997) 67–82
2. Mannila, H., Toivonen, H., Verkamo, A.I.: Discovery of frequent episodes in event sequences. Data Mining and Knowledge Discovery **1** (1997) 259–289
3. Garofalakis, M.N., Rastogi, R., Shim, K.: SPIRIT: Sequential pattern mining with regular expression constraints. In: The VLDB Journal. (1999) 223–234
4. Teahan, W.J., Wen, Y., McNab, R.J., Witten, I.H.: A compression-based algorithm for chinese word segmentation. Computational Linguistics **26** (2000) 375–393
5. Weiss, G.M., Hirsh, H.: Learning to predict rare events in event sequences. In: Knowledge Discovery and Data Mining. (1998) 359–363
6. Magerman, D., Marcus M.: Parsing a natural language using mutual information statistics. In: Proceedings, Eighth National Conference on Artificial Intelligence (AAAI 90). (1990) 984–989
7. Brent, M.R.: An efficient, probabilistically sound algorithm for segmentation and word discovery. Machine Learning **45** (1999) 71–105
8. Ando, R.K., Lee, L.: Mostly-unsupervised statistical segmentation of japanese: Ap-plication to kanji. In: Proceedings of the American Association for Computational Linguistics (NAACL). (2000) 241–248
9. Van Rijsbergen, C.J.: Information Retrieval, 2nd edition. Dept. of Computer Sci-ence, University of Glasgow (1979)

If You Can't See the Pattern, Is It There?

Antony Unwin

Mathematics Institute, University of Augsburg
unwin@math.uni-augsburg.de

Abstract. Analytic methods are capable of finding structure in even complex data sets and, indeed, some methods will find structure whether it is there or not. Confirming and understanding analytic results can be difficult unless some way of visualising them can be found. Both global overviews and displays of local detail are required and these have to be blended intelligently together. This paper discusses the development of coherent graphical tools for exploring and explaining large multidimensional data sets, emphasising the importance of using an interactive approach.

1 Introduction

The value of carrying out visual checks of analytic results is well-known (even if the checks are not carried out as often as they might be). The classic example is regression, where every text clearly recommends that residual plots should be inspected to assess the fit of a model. How they should be inspected is not so clearly stated. Readers get some general, if vague, advice about what they might see if there is a problem. The fit of loglinear models can similarly be investigated by examining the appropriate mosaic plots, but these visual displays are not nearly as widespread.

There are several reasons for examining plots instead of just relying on analytic methods. Plots are easier to grasp in general and easier to explain to data set owners. Background information, which only domain specialists can provide, will only be offered if the specialists can understand what the analyst is trying to do. Plots generally provide a check that the model-fitting procedure is doing what it is supposed to do (although when results are not as expected, it is mostly a problem with the data or with an incorrect specification of the method rather than with the method itself). And plots highlight special features, which are not part of the main data set structure — patterns in the sense of this meeting. There may be a small group of points which is not well fitted or perhaps a group which is exceptionally well fitted. There may be a group which is isolated from the rest. There are many possible structures, which may be observed.

The advantage of visual methods lies in their being complementary to analytic approaches. When an algorithm has detected some patterns of interest, visual methods can be used both to assess their importance and to see if there is anything else interesting in the neighbourhood. It is the first use that is to be discussed in this paper. The second use is more of a supplementary bonus.

D.J. Hand et al. (Eds.): Pattern Detection and Discovery, LNAI 2447, pp. 63–76, 2002.
© Springer-Verlag Berlin Heidelberg 2002

2   Patterns in Large Data Sets

Detecting patterns from plots works very well with small data sets, where only a few plots are needed to display all the information. With large data sets this kind of approach is no longer efficient. Analysts may still spot interesting small-scale features, but it is unlikely that they will find more than a few and some form of guidance is essential to know where to look. This might suggest that visual methods have little part to play in the identification of patterns in large data sets and the task should be left to analytic methods, but that would be the wrong conclusion to draw. Analytic methods will only be able to find patterns for which they have been designed to search. Patterns of other forms will not be found. No algorithm or collection of algorithms can be designed, which can be guaranteed to find all features which might be of interest. Time series demonstrate the difficulty well. Even with a single series it would be necessary to check for single outliers, for groups of outliers, for level change-points, for volatility change-points, for cycles, for trends, for changes in trends, for turning-points, for gaps and so on. All these checks would have to be carried out over all and over sections of the series. With a single short time series visual methods could pick out all of these patterns (if they existed), but this would be impossible for a very long series or for groups of series being analysed in unison.

Large data sets bring new problems for visualisation. In general point displays (where each case is represented by an individual point, as in scatterplots) do not scale up easily. Area-based displays can in prinicple represent any size of data set, but individual details can become lost through screen resolution problems [4]. When only a few cases are selected in a bar the raw display may imply that none are highlighted. The software MANET gets round this by using redmarking to alert the user if such problems arise [12] but that is still not ideal. Some form of zooming is required to show both the pattern and its context. Without context we cannot judge whether a pattern stands out and we cannot evaluate its importance.

Data sets may be considered large because they have many cases or because they have many variables or because of both. When there are many variables in a data set, quite new management problems arise. Scatterplot matrices are effective for up to eight variables, but would be a disaster for forty. Parallel coordinates could manage forty variables quite well, but certainly not one thousand. Sngle displays of large numbers of variables are difficult enough, but handling large numbers of displays of variables is a much bigger problem. They must be monitored, linked and structured. Little consideration seems to have been given to these issues up till now.

3   When Is a Pattern Important?

Patterns may be spurious. If enough random data are examined, there is bound to be something that looks like structure, but it may not mean anything: the infamous paper of Coen, Gomme and Kendall [2] reporting that UK car production led the US Stock market by six quarters being a classic example. Patterns may be "real", but of no practical consequence: the patients diagnosed as having breast cancer are all women. Patterns may be important, if we can be sure they exist. Finding patterns is one activity, possibly performed for large data sets by automatic search algorithms, possibly carried out by individuals exploring the data "by hand". Confirming that patterns both exist and are worthy of note are two further parts of the process.

A pattern can sometimes be assessed statistically by looking at how likely it is that such a result occurs under a reasonable set of assumptions. The difficulties lie in determining what are reasonable assumptions and in deciding how far they can be considered to be satisfied in any given case. Can the data set be assumed to be a random sample from a homogeneous population so that a confidence interval can be constructed for the mean? What about the slews of chi-squared tests that are often carried out and the assumptions necessary for them?

Whether a pattern is important or not finally depends on its context. Is it a curiosity which only applies in an unusual set of circumstances or can it be viewed more generally? Is it merely an unactionable observation or it is a feature which will affect decision-making? The full context can only be established in cooperation with domain experts and this is where a classical difficulty of applying statistics in practice arises. If you tell someone about a pattern, they are liable to be able to explain it in a rational and convincing manner — no matter what it is. This makes it all the more important that any pattern is checked carefully both statistically and visually before being presented.

4   Assessing Detected Patterns Visually

Assessment of patterns takes place at both global and local levels and it is useful to keep these distinct. Global checks look at a pattern in the light of the whole data set. If there are a hundred more interesting patterns in the data set, then there is little point in bothering with number one hundred and one, no matter how interesting it seems to be in its own right. So methods are needed which place individual patterns in the context of the whole data set. Association rules are an excellent example. The TwoKey plot [13] displays all rules found (however many there are) in a single interactive scatterplot of confidence against support. As well as providing an overview in itself, it can be linked to other displays to show how good rules of a particular length or involving similar variables are. Any chosen rule may be considered in comparison to its neighbours in the TwoKey plot by selecting those around it. Additionally lines may be drawn connecting a rule to its logical neighbours (ancestors, descendants or rules with similar variables) to identify the local neighbourhood in variable space.

If a pattern is interesting in a global sense, because it has a relatively high standing, it is still necessary to check if there is not a neighbouring pattern, which is substantially better. With association rules it is quite possible to find a result of the kind $A \rightarrow B$ with both high support and high confidence and yet to discover on closer inspection that $(\text{not}A) \rightarrow B$ is roughly as good. Double Decker plots [6] are an excellent tool for checking individual rules locally.

Patterns which are not major structural features in a data set are multivariate relationships. They are defined by the conditioning variables which describe the area of the data set involved and the variables in which the pattern arises. Multivariate statistical displays are of two kinds, those which use dimension reduction to display the data (for instance, a scatterplot of the data on the first two principal components or an MDS display) and those which stick with the raw data. Certain features may show up in a dimension reduction display, but there is a real risk, especially with patterns, that what is of interest will not show up in a reduced dimension view. Raw data displays are essential for assessing patterns.

5   Visualisation Tools for Patterns

5.1   Patterns Involving Categorical Variables

Statistical methods for multivariate categorical data are less developed than methods for continuous data and results from them are difficult to interpret — how would you explain to someone the fit of a loglinear model with several three-factor interactions? Loglinear models require all cell combinations to be filled and this is unlikely, even with very large data sets, for anything over a few dimensions. Other analytic procedures are required. Until recently there was no effective way of graphically displaying such data, but interactive mosaic plots [5] have extended the possibilities dramatically. Figure 1 shows a fluctuation diagram (a variant of a mosaic plot in which the cells are positioned on a grid) from the Death Sentence data set [8]. The 64 possible combinations of six binary variables (killer's race, victim's race, victim's sex, known/stranger, gun/no gun, robbery/not robbery) are displayed weighted by counts. A weighted histogram of the cell size distribution has been drawn in Figure 2 and the biggest cell selected: 28 cases where a black killed a white male stranger with a gun in a robbery. Observe that there are no cases in the adjacent cell representing the same combination other than that no gun was involved. Interesting also is that there are relatively few cases in the top right quadrant representing cases where a white killed a black.

Fluctuation diagrams are complex displays for representing complex structures and need not only interactive linking, as shown here, to be usable but full interactive features.

5.2   Patterns in Missing Values

Statistics texts tend to ignore missing values. Small data sets don't have them and, especially when there are only a few of them, they are an irritant more than anything else. (Although balanced experimental designs lose all their nice properties when even one value is missing.) Large data sets may have many missing values and methods are needed to deal with them. These depend very much on what structure, if any, is in the patterns of missings.

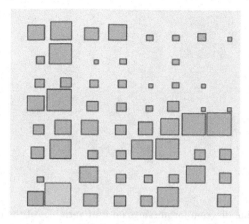

Fig. 1. Fluctuation diagram of six variables from the Death Penalty data set [8]: killer's race, victim's race, victim's sex, known/stranger, gun/no gun, robbery/not robbery

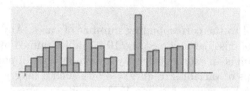

Fig. 2. Weighted histogram of the cell size distribution

Hand et al [3] suggest using a dotplot of cases by variables, where a dot is drawn if the case is missing. This makes for an effective global view for large numbers of variables (in their example there are 45) but not for large numbers of cases (in their example it just about works for 1012 cases, but would clearly have problems with much larger numbers of cases). Patterns of missings across the whole data set are readily visible, but also patterns which only apply to small subgroups of cases. This is a common occurrence in large data sets, where for one reason or another data may be missing on several variables for a particular group of cases.

Unwin et al [12] describe an interactive missing value plot, which displays a bar for each variable. Being an area rather than a point plot, there is in principle no limit to the size of data set which may be displayed. Fully filled bars represent variables with no missings and the empty section of a partially filled bar is proportional to the number of missings on that variable. Interactively selecting the sections representing missings gives some insight into global patterns of missings, but is not nearly informative enough. However, if only a few variables are of interest, say up to 10, then a fluctuation diagram of missings is completely informative. The fluctuation diagram draws a cell for each combination of missing and non-missing amongst the variables and the size of the cell's rectangle is

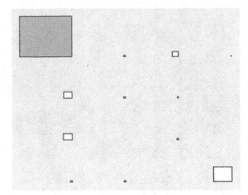

Fig. 3. A fluctuation diagram for missings on four variables of the ultrasound dataset. 82% of the cases are complete (cell top left) and 11% are missing on all four variables. Another 6% are missing only on a single variable. Several possible combinations of missings do not occur at all.

made proportional to the corresponding number of cases. Any subgroup with a common pattern of missings amongst the 10 variables immediately stands out. A fluctuation diagram is an excellent display for evaluations of patterns of missings amongst subgroups of variables. Figure 3 shows such a diagram for the pattern of missings amongst four variables for 4571 hospital pregnancies. The variables were three baby size measurements estimated from ultrasound images taken in the days before delivery and the number of days before pregnancy when the ultrasound was carried out.

If there is a large proportion of complete cases, that cell may mask the others. An option to limit the size of this cell is under consideration.

5.3   Geographic Patterns

Visual tools should not be viewed solely as a means of checking analytically derived results. In some kinds of data set it is extremely difficult to achieve any convincing analytic results. Geographic data are a prime example because there are so many special geographic features that need to be taken account of in any analysis: national boundaries, mountain ranges, rivers, transport routes and so on. The units of analysis may be areas of completely different sizes and shapes. Cities are usually small areas with large populations and consequently high population density, while there are huge tracts of land in rural areas where few people live. Just consider the United States of America.

Visual displays are an essential tool for detecting geographic patterns not just because analytic tools fail, but because people have a great deal of experience of looking at maps. They are good at recognising features and interpreting them in context. Figure 4 shows a map of electoral constituencies in Germany after the 1998 election with those constituencies highlighted where the PDS (a descendant

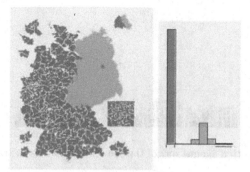

Fig. 4. The histogram on the right shows the % PDS votes by constituency for the Bundestag election in 1998. The map on the left shows the constituencies with major cities and the Ruhr magnified. Berlin is magnified to the top right of the map, but also appears to scale in the centre of East Germany.

of the former East German Communist party) did relatively well. Any one with knowledge of German geography immediately recognises that the PDS did well only in the East and, in Berlin, only in East Berlin.

6   Visual Tools for Elucidating Displays of Patterns — Interaction

Graphic displays were not very popular in the early days of statistical computing because the screen representations and the print quality were so poor. Nowadays high-quality presentation displays can be produced very easily (even if it is more common to produce poor quality ones). Presentation displays do not fully exploit the power of modern computing, as anyone who surfs the web will be aware. Graphics have to be interactive it they are to encourage the effective exploration of data sets. [11] discusses the essential interactive requirements for a statistical graphic and all are relevant to the assessment of detected patterns.

Querying is valuable both to supply comprehensive information about a pattern and to explore its context. Interaction is crucial. Consider the example of examining a subset of points in a scatterplot. Even if the points were sufficiently spread out that they could be labelled, information on other variable values is usually necessary to complete the picture. Furthermore any permanent labelling affects the appreciation of the points' position in relation to the rest of the scatterplot. For a fluctuation diagram like that shown in Figure 1 querying is essential to even begin to grasp the information displayed.

Any display of categorical variables is affected by the order of the variables and by the ordering of the categories within the variables. Sorting tools, either manual or based on statistics or selections, are necessary to check the sensitivity of a pattern. Changing the order of the variables in Figure 1 would produce

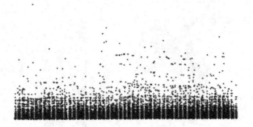

Fig. 5. Amount plotted against day of the year for 277000 bank transactions between firms.

a different picture and hence a different view of the relationships between the variables.

As data sets become larger, the numbers of categories in categorical variables tend to increase. More unusual cases arise and are recorded. This can lead to both tables and displays becoming excessively large and unclear. In a data set of political opinion polls in Germany over 20 years there was data for support for 12 different responses, including sometimes treating the main conservative party as one response and sometimes as two separate response (CDU/CSU). Removing all categories below a certain size or combining all the smallest categories may often suffice, but would be a risky policy in general. Sometimes it is one of the smaller categories that is of most interest because the pattern for that group is so different from the rest or from a related subgroup of the rest. Whatever the appropriate comparison, we need to be able to work with the graphic in an intuitive way to achieve it. Combining selected categories, conditioning on a subset of categories, reordering categories are all basic data manipulation tools in this situation. While all these tasks may be carried out by one or other form of variable transformation, they are all messy and time-consuming if carried out in this way (just think of renaming categories to force a reordering compared to reordering interactively).

It goes without saying in a discussion of patterns in small subsets that it can be important to zoom into part of a display. Deciding exactly where and how far to zoom in depends on what becomes visible, so that being able to vary this on the fly is very useful. Having to set limits in a dialog box with subsequent redrawing of the screen is an unsatisfactory means of checking details. Interactivity provides the flexibility needed. In Figure 5 of 277000 large-scale bank transactions showing amount against day of the year there is small subgroup of similar high values in the middle of the year. Figure 6 shows that part of the plot after an interactive zooming and rescaling. It is now clear that there are possibly two groups of different levels and further zooming in and linking with other variables is the next step. (It turned out that the transactions did form a uniform group, involving the same currency, the same type of transaction and the same source.)

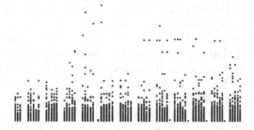

Fig. 6. Amount plotted against day of the year for May to July.

Zooming might be defined by visual selection or by any other kind of selection. Data Desk [14] offers a clever tool called hot selection for this. Any output which is assigned to be influenced by selection is automatically updated to only display the currently selected points and is also rescaled accordingly. In the last example this means that when you link from the small group of eight points back to the barchart of currencies you can see which currency was involved. A barchart of eight points is trivial to interpret, finding eight highlighted points in a barchart of almost 300000 points is impossible. The power of hot selection lies in the concept and in its implementation. Updating is almost instantaneous for such a large data set, even when several outputs are updated simultaneously.

7 Coherent Graphical Tools – Integrating Global and Local Views

7.1 Protocols

Analysing large data sets requires good organisation and management. It is all too easy to generate hundreds of tables and graphics, which swamp any insights that might be contained within them. Many systems routinely store history protocols of what actions have been performed. Amongst statistics software packages Data Desk generates a sequence of icons and S-plus a more conventional command and output listing. Neither is easy to read and trying to retrace your steps even a short time later is difficult. Tree structures have been suggested to display the analysis paths followed, but in exploratory analyses these are often confusing because of false trails and repetitions. The ideal would be some kind of edited history supported by a tree browser (alternative tree browers are discussed in Shneiderman [9]), though Peter Huber has commented that editing protocols is far more trouble than it turns out to be worth. For protocols to be useful they have to be more than just bookkeeping.

Indexing results automatically when they are produced is an effective means of doing this and yet it is surprising how rarely it is done. The index should offer interactive linking and guidance to its contents. For example, in a related context Stephan Lauer's software TURNER for exploring contingency tables [7] sets up a list of all models fitted. Models can be queried from the list and pairs of models can be directly compared, where nesting permits. One of the design issues which arose in TURNER was how the fitted models should be stored, but it became apparent that that was not necessary. If more details of any model were needed than were carried in the index list, it was just as quick (and simpler) to recalculate the model. This will be true for many models and is a good general principle to follow; output files are smaller and much more manageable.

It is obvious that when recalculation is expensive that this approach cannot be used, but there is another situation in whic h recalculating on demand will not work. Classical statistical methodology assumes that the data set is fixed. If analyses are being made of a database and the data are left there, it is quite possible for results to change between two calculations if the database is continually being updated. Although this is not a major problem, it can cause difficulties and expecting the user to be aware of it and take the required measures to preserve consistency is unrealistic. Time stamping displays (a feature which Data Desk offers) is a step in the right direction.

Linking large numbers of tables and graphics is rather like designing websites. Unless it is very carefully planned and carried out, there is a good chance that you end up with an untidy jumble as the structure grows in a higgledy piggledy fashion. But you can't fully plan data analyses, as the paths that are followed will depend on the results which are found. So we should distinguish between the kind of structuring that is needed during the analysis process and the kind of structuring that can be achieved for a given collection of results. The former requires the automatic indexing that has already been mentioned and a protocol of the paths followed, while the latter is much more of an information design problem, indeed quite like designing a website. Although one would think that output design would be an important part of any software, it is interesting to note that data mining software may be found with impressive input design and yet almost retro-design of output. Users of commercial packages from the sixties would feel quite at home examining some of the outputs delivered.

7.2   An Overview and Action Window

Monitoring large numbers of graphics is possible if they are linked. In the PISSARRO software, which is currently under development in Augsburg, querying an individual association rule in a TwoKey plot of all rules generates a Double Decker plot of the variables involved for local inspection. In geographic software (and in games like Sim City) it is common to display a miniature version of the whole window when zooming to indicate where the zoom has taken place. This is an effective way of keeping track of where you are and is just as relevant for scatterplots as for maps. The principle is clear: not only should it be possible to move from a global view to a local view to check and explore details, but it must also be possible to move back from the local to the global. While this sounds straightforward and rather elementary, it can be difficult to implement in practice because there is not one global level or one local level, there is potentially a whole range of levels. Is a view currently local (a subset of a bigger view) or global (providing context to one or more local views)? And what does local mean? Which variables define the conditioning? If appropriate comparisons of possibly interesting findings are to be made, these questions are very important. Results have to have a high ranking on a global view and stand out locally to be worth considering.

The keys to the overview and control of many views are:
a) orientation
b) monitoring
c) navigation
d) focussing
e) comparisons.

a) Orientation means that the system assists the user to know where he is at the moment (both within the data set, in data space, and within the output displays, in results space). Knowing where you are includes not just a (preferably) exact definition but also an awareness of context.

b) Monitoring implies that the user can keep track of what is going on, looking back and looking forward.

c) Navigation tools help you get around. How did you arrive at this point? Where could you go next? How can you get to where you want to go to?

d) Focussing involves more than zooming in or drilling down. The displays should certainly concentrate on the subset of interest, but not exclusively. The user needs a detailed view within a broader context.

e) Comparisons are essential in a scientific study. Systems should support the identification, carrying out and reporting of relevant comparisons. This is a challenging task in large, comlex data sets.

Central to a successful implementation of all of these aspects is conditioning, describing where you are at the moment in terms of variable definitions in an intuitive and flexible way.

Trellis diagrams [1] offer one way of presenting displays of many data subsets and are in one sense an example of an overview window, though not an action one. Data displays are drawn for subsets of the data for all possible combinations of a number of conditioning variables (which may be categorical or discretised continuous variables). The basic data display may be for one or two continuous variables (usually scatterplots, but with one variable dotplots, boxplots, or histograms would be possible). For a categorical panel variable barcharts or piecharts might be drawn. All panels are the same physical size. The structure is fairly clear and a lot of thought has gone into how labelling should be implemented. The disadvantages are that all the displays are shown at once, which complicates focussing, there is no interaction and, in particular, there is no possibility of reordering the variables or changing the conditioning in any way other than by drawing another trellis diagram. In principle interaction could be added and this would emphasise the similarities between trellis diagrams and mosaic plots which we discuss now.

Mosaic plots, although associated only with categorical data, actually have a structure with similarities to that of trellis diagrams. All variables are used to condition the display and the counts of the resulting combinations are represented by a rectangle of an appropriate size. There is no real difference in the way conditioning is carried out, the difference lies in the final panels. Closer similarity can be seen if a multiple bar chart variant of the mosaic plot is used. This would be like a trellis diagram with a panel barchart display, except that all the trellis barcharts would occupy equal-sized panels while the barcharts in the mosaic plot would be of different sizes corresponding to the numbers in that data subset.

It is often the case that related ideas are found under quite different names and it is worth noting that displays known as "small multiples", where the same form of display is used for many different subsets or individuals in a matrix layout, also have properties in common with trellis diagrams and mosaics. What distinguishs mosaics from the other two most is that interactive implementations exist. We can see using these how windows with structures very like mosaic plots could be used to both survey complex data structures and to provide action tools for exploring them.

Interactive mosaic plots are available in MANET [5] and in Mondrian [10]. The MANET implementation provides many interactive features which could be adopted for an overview and action window — querying individual objects, moving through dimensions using the arrow keys (aggregating and disaggregating), reordering variables with drag and drop, choosing other representation forms, reordering categories using separate linked barcharts. What is missing is a focussing tool, but that could be added by implementing a tool like Data Desk's hot selection [14]. If a graphic is linked to a hot selector then it is automatically limited to the cases in the currently selected subset. As the scale is linked by default as well, this means that the relevant display expands to fill the window. The implementation of hot selection in Data Desk is very impressive, as displays are instanteously refocussed as the selection is changed. What is missing is an orientation window or global view to show what the selection is in the data set and what proportion of the data set is selected.

Providing orientation in a cartographic view is straightforward, we just have to mark on the global map where the focus of our display window lies. Maps are readily interpretable and viewers will know what location is meant. The same cannot be said of viewers of mosaic plots or trellis diagrams. There are no natural and familiar boundaries to guide the eye and there will be a different overview for each data set. An approach is needed which both provides an overview and labels it effectively. This could be achieved by marking the focus in the overview window, lightening the non-focussed areas and adding a text description of the relevant conditioning. The Augsburg software CASSATT for interactive parallel coordinates [15] uses this combination of highlighting selected cases but optionally lightening or "ghosting" non-selected cases and this works very well in de-emphasising the cases that are not currently of interest while still giving context. It is more effective than just highlighting the selected cases and the same effect would be anticipated for focussing. Text descriptions are useful for exact specifications, but do not give context. One solution might be to additionally provide univariate plots of the conditioning variables with the selections highlighted. For the occasional user (and probably the experienced user as well) this will work better if the variables are always represented by the same plots and are always in the same screen position, but this will only be possible for a limited number of variables (even if highlighting and ghosting are used for these displays as well).

Although the suggestions made above already imply a considerable overhead for a system, they are still incomplete. Focussing on a subset will almost certainly require making comparisons with other subsets, usually with "local" subsets. Action controls are needed to identify relevant neighbours and facilitate comparisons, either in parallel or sequentially.

8 Conclusions

It is easy to find many patterns in a large data set, but difficult to decide if they have any statistical support and difficult to evaluate their importance. Visual methods complement analytic procedures and offer an alternative approach. Interaction is essential both to interpret displays and to link them together. Both global and local views are necessary, but they should be applied in an integrated structure rather than just as a collection of individual displays. Support for the user must include orientation, monitoring, navigation, focussing and comparisons. A new variant of mosaic plots could provide the overview and action window, which is needed, especially if hot selection is added to the range of interactive features and if effective conditioning methods can be designed and implemented.

References

1. Becker, R., Cleveland, W.S., Shyu, M-J.: The Visual Design and Control of Trellis Display Journal of Computational and Graphical Statistics, 5, 123–155 (1996).
2. Coen, P. G., Gomme, E.D. and Kendall, M.G.: Lagged Relationships in Economic Forecasting JRSS A, 132, (1969)133–163.
3. Hand, D. J., Blunt, G., Kelly, M.G., Adams, N.M.: Data Mining for Fun and Profit Statistical Science, 15(2), (2000)111–131.
4. Hofmann, H.: Graphical Stability of Data Analysing Software, In Klar,R., Opitz, O. (Eds.) Classification and Knowledge Organisation, Freiburg Germany (1997) 36–43.
5. Hofmann, H.: Exploring categorical data: interactive mosaic plots Metrika, 51(1), (2000)11–26.

6. Hofmann, H., Wilhelm, A.: Visual Comparison of Association Rules, Computational Statistics, 16, (2001) 399–415.
7. Lauer, S.: Turning the Tables with Turner, SCGN, 11(1), (2000) 5–9.
8. Morton, S. C., Rolph, J. E.: Racial Bias in Death Sentencing: Assessing the Statistical Evidence, In Morton, S. C., Rolph, J. E. (Eds.) Public Policy and Statistics: Case Studies from RAND, New York: Springer (2000) 94–116.
9. Shneiderman, B.: Designing the User Interface. Addison-Wesley (1998).
10. Theus, M.: Mondrian www1.math.uni-augsburg.de/Mondrian/. In Augsburg: Rosuda (2002)
11. Unwin, A.: Requirements for Interactive Graphics Software for Exploratory Data Analysis, Computational Statistics, 14, (1999) 7–22.
12. Unwin, A. R., Hawkins, G., Hofmann, H., and Siegl, B.: Interactive Graphics for Data Sets with Missing Values – MANET Journal of Computational and Graphical Statistics, 5,(1996) 113–122.
13. Unwin, A., Hofmann, H., Bernt, K.: The TwoKey Plot for Multiple Association Rules Control, In De Raedt, L. Siebes, A. (Eds.) Proc. of the PKDD/ECML, Freiburg Germany (2001) 472–483.
14. Velleman, P.: Data Desk 6.1 Data Description (1999)
15. Winkler, S.: CASSATT www1.math.uni-augsburg.de/Cassatt/. In Augsburg: Rosuda (2000)

Dataset Filtering Techniques in Constraint-Based Frequent Pattern Mining

Marek Wojciechowski and Maciej Zakrzewicz

Poznan University of Technology
Institute of Computing Science
ul. Piotrowo 3a, 60-965 Poznan, Poland
Marek.Wojciechowski@cs.put.poznan.pl
Maciej.Zakrzewicz@cs.put.poznan.pl

Abstract. Many data mining techniques consist in discovering patterns frequently occurring in the source dataset. Typically, the goal is to discover all the patterns whose frequency in the dataset exceeds a user-specified threshold. However, very often users want to restrict the set of patterns to be discovered by adding extra constraints on the structure of patterns. Data mining systems should be able to exploit such constraints to speed-up the mining process. In this paper, we focus on improving the efficiency of constraint-based frequent pattern mining by using dataset filtering techniques. Dataset filtering conceptually transforms a given data mining task into an equivalent one operating on a smaller dataset. We present transformation rules for various classes of patterns: itemsets, association rules, and sequential patterns, and discuss implementation issues regarding integration of dataset filtering with well-known pattern discovery algorithms.

1 Introduction

Many data mining techniques consist in discovering patterns frequently occurring in the source dataset. The two most prominent classes of patterns are frequent itemsets [1] and sequential patterns [3]. Informally, frequent itemsets are subsets frequently occurring in a collection of sets of items, and sequential patterns are the most frequently occurring subsequences in sequences of sets of items. Frequent itemsets themselves provide useful information on the correlations between items in the database. Nevertheless, discovered frequent itemsets are very often treated only as the basis for association rule generation [1]. Frequent itemsets, association rules, and sequential patterns were introduced in the context of market basket analysis but their applications also include fraud detection, analysis of telecommunication systems, medical records, web server logs, etc.

Typically, in frequent pattern mining the goal is to discover all patterns whose frequency (called support) in the source dataset exceeds a user-specified threshold. If frequent patterns discovered are to be used to generate rules, the minimum accepted confidence of a rule also has to be specified. Additionally, in sequential pattern discovery several time constraints have been proposed to be

D.J. Hand et al. (Eds.): Pattern Detection and Discovery, LNAI 2447, pp. 77–91, 2002.
© Springer-Verlag Berlin Heidelberg 2002

used when deciding if a given pattern is contained in a given sequence from the source dataset [10]. Since it was shown that derivation of rules from patterns is a straightforward task, the research focused mainly on improving the efficiency of algorithms discovering all patterns whose support in the source dataset exceeds a user-specified threshold.

However, it has been observed that users are very often interested in patterns that satisfy more sophisticated criteria, for example concerning size, length, or contents of patterns. Data mining tasks involving the specification of various types of constraints can be regarded as data mining queries [7]. It is obvious that additional constraints regarding the structure of patterns can be verified in a post-processing step, after all patterns exceeding a given minimum support threshold have been discovered. Nevertheless, such a solution cannot be considered satisfactory since users providing advanced pattern selection criteria may expect that the data mining system will exploit them in the mining process to improve performance. In other words, the system should concentrate on patterns that are interesting from the user's point of view, rather than waste time on discovering patterns the user has not asked for [5].

We claim that techniques applicable to constraint-driven pattern discovery can be classified into the following groups:

1. post-processing (filtering out patterns that do not satisfy user-specified pattern constraints after the actual discovery process);
2. pattern filtering (integration of pattern constraints into the actual mining process in order to generate only patterns satisfying the constraints);
3. dataset filtering (restricting the source dataset to objects that can possibly contain patterns that satisfy pattern constraints).

As the post-processing solution was considered unsatisfactory, the researchers focused on incorporating pattern constraints into classic pattern discovery algorithms. This led to the introduction of numerous constraint-based pattern discovery methods (e.g. [4][11]), all of which fall into the second group according to our classification. It should be noted that some of the methods from this class generate a superset of the collection of patterns requested by user, which means that a post-processing phase might still be required. Nevertheless, all these methods use pattern constraints to reduce the number of generated patterns, leading to a smaller set of patterns to be verified in the post-processing step than in case of classic algorithms.

In this paper we discuss an alternative approach to constraint-based pattern discovery, called *dataset filtering*. Dataset filtering is based on the observation that for some classes of pattern constraints, patterns satisfying them can only be contained in objects satisfying the same or similar constraints. The key issue in dataset filtering is derivation of filtering predicates to be applied to the source dataset from pattern constraints specified by a user. Dataset filtering is a general technique applicable to various types of patterns but for a particular class of patterns and a given constraint model distinct derivation rules have to provided. We focus on two types of patterns: frequent itemsets and sequential patterns. We also discuss extensions required to handle association rules. We assume a relatively simple constraint model with pattern constraints referring to the size or length of patterns or to the presence of a certain subset or subsequence. Nevertheless, we believe that types of constraints we consider are the most intuitive and useful in practice.

Conceptually, dataset filtering transforms a given data mining task into an equivalent one operating on a smaller dataset. Thus, it can be integrated with any pattern discovery algorithm, possibly exploiting other constraint-based pattern discovery techniques. In this paper we focus on the integration of dataset filtering techniques within the *Apriori* framework. We discuss possible implementations of dataset filtering within *Apriori*-like algorithms, evaluating their strengths and weaknesses.

2   Background and Related Work

2.1   Frequent Itemsets and Association Rules

Let $L = l_1, l_2, ..., l_m$ be a set of literals, called *items*. An *itemset* X is a non-empty set of items $(X \subseteq L)$. The *size* of an itemset X is the number of items in X. Let D be a set of variable size itemsets, where each itemset T in D has a unique identifier and is called a *transaction*. We say that a transaction T *contains* an item $x \in L$ if x is in T. We say that a transaction T *contains* an itemset $X \subseteq L$ if T contains every item in the set X. The *support* of the itemset X is the percentage of transactions in D that contain X. The problem of mining frequent itemsets in D consists in discovering all itemsets whose support is above a user-defined support threshold.

An *association rule* is an implication of the form $X \rightarrow Y$, where $X \subseteq L$, $Y \subseteq L$, $X \cap Y = \emptyset$. We call X the *body* of a rule and Y the *head* of a rule. The *support* of the rule $X \rightarrow Y$ in D is the support of the itemset $X \cup Y$. The *confidence* of the rule $X \rightarrow Y$ is the percentage of transactions in D containing X that also contain Y. The problem of mining association rules in D consists in discovering all association rules whose support and confidence are above user-defined minimum support and minimum confidence thresholds.

2.2   Sequential Patterns

Let $L = l_1, l_2, ..., l_m$ be a set of literals called *items*. An *itemset* is a non-empty set of items. A *sequence* is an ordered list of itemsets and is denoted as $< X_1 X_2...X_n >$, where X_i is an itemset $(X_i \subseteq L)$. X_i is called an *element* of the sequence. The *size* of a sequence is the number of items in the sequence. The *length* of a sequence is the number of elements in the sequence.

We say that a sequence $X =< X_1 X_2...X_n >$ is a *subsequence* of a sequence $Y =< Y_1 Y_2...Y_m >$ if there exist integers $i_1 < i_2 < ... < i_n$ such that $X_1 \subseteq Y_{i_1}, X_2 \subseteq Y_{i_2}, ..., X_n \subseteq Y_{i_n}$. We call $< Y_{i_1} Y_{i_2}...Y_{i_n} >$ an *occurrence* of X in Y.

Given a sequence $Y =< Y_1 Y_2...Y_m >$ and a subsequence X, X is a *contiguous* subsequence of Y if any of the following conditions hold: 1) X is derived from Y

by dropping an item from either Y_1 or Y_m. 2) X is derived from Y by dropping an item from an element Y_i which has at least 2 items. 3) X is a contiguous subsequence of X', and X' is a contiguous subsequence of Y.

Let D be a set of variable length sequences (called *data-sequences*), where for each sequence $S = < S_1 S_2 ... S_n >$, a timestamp is associated with each S_i. With no time constraints we say that a sequence X is *contained* in a data-sequence S if X is a subsequence of S. We consider the following user-specified time constraints while looking for occurrences of a given sequence in a given data-sequence: minimal and maximal gap allowed between consecutive elements of an occurrence of the sequence (called *min-gap* and *max-gap*), and time window that allows a group of consecutive elements of a data-sequence to be merged and treated as a single element as long as their timestamps are within the user-specified *window-size*.

The *support* of a sequence $< X_1 X_2 ... X_n >$ in D is the fraction of data-sequences in D that contain the sequence. A *sequential pattern* (also called a *frequent sequence*) is a sequence whose support in D is above the user-specified minimum support threshold.

2.3   Review of Classic Pattern Mining Algorithms

The majority of frequent itemset and sequential pattern discovery algorithms fall into two classes: *Apriori*-like methods and *pattern-growth* methods. The first group of methods is based on the *Apriori* algorithm for frequent itemset mining [2]. *Apriori* relies on the property (called *Apriori* property) that an itemset can only be frequent if all of its subsets are frequent. It leads to a level-wise procedure. First, all possible 1-itemsets (itemsets containing 1 item) are counted in the database to determine frequent 1-itemsets. Then, frequent 1-itemsets are combined to form potentially frequent 2-itemsets, called candidate 2-itemsets. Candidate 2-itemsets are counted in the database to determine frequent 2-itemsets. The procedure is continued until in a certain iteration none of the candidates turns out to be frequent or the set of generated candidates is empty. Several extensions were added to improve the performance of *Apriori* (e.g. by reducing the number of database passes). The algorithm also served as a basis for algorithms discovering other types of patterns including sequential patterns.

The most prominent sequential pattern discovery algorithm from the Apriori family is *GSP*, introduced in [10]. *GSP* exploits a variation of the Apriori property: all contiguous subsequences of a frequent sequence also have to be frequent. In each iteration, candidate sequences, are generated from the frequent sequences found in the previous pass, and then verified in a database scan. It should be noted that *GSP* (and its variants) is the only sequential pattern discovery algorithm capable of handling time constraints (max-gap, min-gap, and window-size).

Recently, a new family of pattern discovery algorithms, called pattern-growth methods (see [6] for a review), has been developed for discovery of frequent patterns. The methods project databases based on the currently discovered frequent patterns and grow such patterns to longer ones in corresponding projected databases. Pattern-growth methods are supposed to perform better than *Apriori*-like algorithms in case of low minimum support thresholds. Nevertheless, practical studies [12] show that for real datasets *Apriori* (or its variants) might still be a more efficient solution. Moreover, in the context of sequential patterns pattern-growth methods still do not offer full functionality of *GSP*, as they do not handle time constraints.

2.4   Previous Work on Constraint-Based Pattern Mining

As we mentioned earlier the research on constraint-based pattern mining focused on incorporating pattern constraints into classic pattern discovery algorithms, especially in the context of frequent itemsets and association rules. Pattern constraints in frequent itemset and association rule mining were first discussed in [11]. Constraints considered there had a form of a Boolean expression in the disjunctive normal form built from elementary predicates requiring that a certain item is or is not present. The algorithms presented were *Apriori* variants using sophisticated candidate generation techniques. Rule constraints were handled by transforming them into itemset constraints. It was observed that after discovering all itemsets that can be used to generate the rules of interest, one extra scan of the dataset is required to count the supports of some subsets, required to evaluate confidences of some rules, and not known since the subsets did not satisfy the derived itemset constraints.

In [8], two interesting classes of itemset constraints were introduced: anti-monotonicity and succinctness, and methods of handling constraints belonging to these classes within the *Apriori* framework were presented. The methods for succinct constraints again consisted in modifying the candidate generation procedure. For anti-monotone constraints it was observed that in fact almost no changes to *Apriori* are required to handle them. A constraint is anti-monotone if the fact that an itemset satisfies it, implies that all of its subsets have to satisfy the constraint too. The minimum support threshold is an example of an anti-monotone constraint, and any extra constraints of that class can be used together with it in candidate pruning.

In [9], constraint-based discovery of frequent itemsets was analyzed in the context of pattern-growth methodology. In the paper, further classes of constraints were introduced, some of which could not be incorporated into the *Apriori* framework.

On the other hand, very little work concerning constraint-driven sequential pattern discovery has been done so far. In fact, only the algorithms from the *SPIRIT* family [4] exploit pattern structure constraints in order to improve performance. These algorithms can be seen as extensions of *GSP* using advanced candidate generation and pruning techniques. In the *SPIRIT* framework, pattern constraints are specified as regular expressions, which is an especially convenient method if a user wants to significantly restrict the structure of patterns to be discovered. It has been shown experimentally that pushing regular expression constraints deep into the mining process can reduce processing time by more than an order of magnitude.

3   Dataset Filtering in Constraint-Based Pattern Mining

In constraint-based pattern mining, we identify the following classes of constraints: database constraints, statistical constraints, pattern constraints, and time constraints. Database constraints are used to specify the source dataset. Statistical constraints are used to specify thresholds for the support and confidence measures. Pattern constraints specify which of the frequent patterns are interesting and should be returned by the query. Finally, time constraints used in sequential pattern mining influence the process of checking whether a given data-sequence contains a given pattern.

Basic formulations of pattern discovery problems do not consider pattern constraints. We model pattern constraints as a conjunction of basic Boolean predicates referring to pattern size or length (*size constraints*) or regarding the presence of a certain subset or subsequence (*item constraints*).

It should be noted that not all pattern predicates support the dataset filtering paradigm. For example, if a user is looking for frequent itemsets whose size exceeds a given threshold, it is rather obvious that such itemsets can be contained only in transactions whose size exceeds the same threshold. Thus, smaller transactions can be excluded from the mining process, which should lead to performance gains. On the other hand, if a user is interested in itemsets having the size not exceeding a given threshold, dataset filtering is not applicable as such itemsets can be contained in any transaction.

For each of the pattern predicates present in pattern constraints supporting dataset filtering, the corresponding predicate on transactions or data-sequences has to be derived. The resulting dataset filtering predicate is formed as a conjunction of those derived predicates (recall that we consider pattern constraints having the form of a conjunction of pattern predicates). The filtering predicate is then used to discard objects in the source dataset that cannot contain the patterns of interest. Below we identify pattern predicates in case of which dataset filtering is applicable in the context of frequent itemsets, association rules and sequential patterns. For each of the identified pattern predicates we provide a corresponding predicate on source objects, to be used in dataset filtering. As we will show, dataset filtering is rather straightforward in frequent itemset discovery but becomes more complicated in sequential pattern mining thanks to time constraints.

3.1   Dataset Filtering in Frequent Itemset Discovery

Let us consider the following predicate types that can appear in pattern constraints of a frequent itemset query:

- $\rho(\mathbf{SG}, \alpha, itemset)$ - true if itemset size is greater than α, false otherwise;
- $\rho(\mathbf{C}, \gamma, itemset)$ - true if γ is a subset of the itemset, false otherwise;

Theorem 1. *Itemsets of size greater than k cannot be contained in a transaction whose size is not greater than k.*

Proof. The proof is obvious since an itemset is contained in a transaction if it is a subset of the set of items present in the transaction.

Theorem 2. *Frequent itemsets, to be returned by a data mining query, containing a given subset can be supported only by transactions containing that subset.*

Proof. An itemset is contained in a transaction if it is a subset of the set of items present in the transaction. The transitivity of set inclusion relationship implies that if an itemset having a given subset is contained in a given transaction, the transaction also has to contain the subset.

According to the above theorems the following dataset filtering predicates are applicable in frequent itemset mining:

- $\tau(\mathbf{SG}, \alpha, transaction)$ - true if the size of the transaction is greater than α, false otherwise;
- $\tau(\mathbf{C}, \gamma, transaction)$ - true if the transaction contains the set γ, false otherwise;

For a frequent itemset query with pattern constraints, an appropriate dataset filtering predicate is derived in the following way: For each of the pattern predicates from the left column of Table 1 present in the query, the corresponding transaction predicate from the right table column is added to the dataset filtering predicate.

Table 1. Derivation rules for frequent itemset mining

Pattern predicate	Transaction predicate
$\rho(\mathbf{SG}, \alpha, itemset)$	$\tau(\mathbf{SG}, \alpha, transaction)$
$\rho(\mathbf{C}, \gamma, itemset)$	$\tau(\mathbf{C}, \gamma, transaction)$

3.2   Dataset Filtering in Association Rule Discovery

Let us consider the following predicate types that can appear in pattern constraints of an association rule query:

- $\rho(\mathbf{SG}, \alpha, rule)$ - true if the number of items in the rule is greater than α, false otherwise;
- $\rho(\mathbf{C}, \gamma, rule)$ - true if all items from γ are present in the rule, false otherwise;
- $\rho(\mathbf{SG}, \alpha, body(rule))$ - true if the size of the rule's body is greater than α, false otherwise;
- $\rho(\mathbf{C}, \gamma, body(rule))$ - true if γ is a subset of the rule's body, false otherwise;
- $\rho(\mathbf{SE}, \alpha, body(rule))$ - true if the size of the rule's body is α, false otherwise;

- $\rho(\mathbf{E}, \gamma, body(rule))$ - true if the rule's body is equal to γ, false otherwise;
- $\rho(\mathbf{SG}, \alpha, head(rule))$ - true if the size of the rule's head is greater than α, false otherwise;
- $\rho(\mathbf{C}, \gamma, head(rule))$ - true if γ is a subset of the rule's head, false otherwise;
- $\rho(\mathbf{SE}, \alpha, head(rule))$ - true if the size of the rule's head is α, false otherwise;
- $\rho(\mathbf{E}, \gamma, head(rule))$ - true if the rule's head is equal to γ, false otherwise;

We observe that the rule predicates can be directly transformed into predicates on itemsets that can be used to generate the rules having the desired properties. All the items required in the rule, rule's body, or rule's head have to appear in the frequent itemsets from which the rules are to be generated. The size threshold on rules implies the same threshold on itemsets. However, if a given size of the rule's body or head is required, then the itemset must have at least one more item (neither the head nor the body can be empty). Table 2 presents a corresponding itemset predicate for each of the rule predicates. (In fact, if the predicates for both head and body are present, and at least one of them is a size predicate, a more restrictive itemset size predicate can be derived. We omit the details for the sake of simplicity.)

Table 2. Rule predicates and their corresponding itemset predicates

Rule predicate	Itemset predicate
$\rho(\mathbf{SG}, \alpha, rule)$	$\rho(\mathbf{SG}, \alpha, itemset)$
$\rho(\mathbf{C}, \gamma, rule)$	$\rho(\mathbf{C}, \gamma, itemset)$
$\rho(\mathbf{SG}, \alpha, body(rule))$	$\rho(\mathbf{SG}, \alpha + 1, itemset)$
$\rho(\mathbf{C}, \gamma, body(rule))$	$\rho(\mathbf{C}, \gamma, itemset)$
$\rho(\mathbf{SE}, \alpha, body(rule))$	$\rho(\mathbf{SG}, \alpha, itemset)$
$\rho(\mathbf{E}, \gamma, body(rule))$	$\rho(\mathbf{C}, \gamma, itemset)$
$\rho(\mathbf{SG}, \alpha, head(rule))$	$\rho(\mathbf{SG}, \alpha + 1, itemset)$
$\rho(\mathbf{C}, \gamma, head(rule))$	$\rho(\mathbf{C}, \gamma, itemset)$
$\rho(\mathbf{SE}, \alpha, head(rule))$	$\rho(\mathbf{SG}, \alpha, itemset)$
$\rho(\mathbf{E}, \gamma, head(rule))$	$\rho(\mathbf{C}, \gamma, itemset)$

As the rule predicates implicate itemset predicates, there is no need to provide separate derivation rules for dataset filtering predicates to be used in association rule mining. It should be noted that while the collection of discovered frequent itemsets supporting the constraints derived from the rule constraints is sufficient to generate all the required rules, it may not contain all the itemsets needed to evaluate confidences of the rules as certain subsets of those itemsets may not satisfy pattern constraints. This problem is not specific to our dataset filtering techniques, and has to be solved by an extra scan of the dataset. In that extra scan, supports of itemsets whose support is needed but not known have to be counted.

3.3   Dataset Filtering in Sequential Pattern Discovery

Let us consider the following predicate types that can appear in pattern constraints of a sequential pattern query:

- $\pi(\textbf{SG}, \alpha, pattern)$ - true if pattern size is greater than α, false otherwise;
- $\pi(\textbf{LG}, \alpha, pattern)$ - true if pattern length is greater than α, false otherwise;
- $\pi(\textbf{C}, \beta, pattern)$ - true if β is a subsequence of the pattern, false otherwise;
- $\rho(\textbf{SG}, \alpha, pattern_n)$ - true if the size of the n-th element of the pattern is greater than α, false otherwise;
- $\rho(\textbf{C}, \gamma, pattern_n)$ - true if γ is a subset of the n-th element of the pattern, false otherwise;

Theorem 3. *Sequential patterns of size greater than k cannot be contained in a data-sequence whose size is not greater than k.*

Proof. The proof is obvious since an occurrence of a pattern in a sequence must consist of the same number of items as the pattern.

Theorem 4. *Sequential patterns of length greater than k, to be returned by a data mining query, can be contained only in data-sequences which contain some sequence of length $k + 1$ using max-gap, min-gap, and window-size specified in the query.*

Proof. Each sequential pattern of length greater than k has at least one contiguous subsequence of length $k + 1$. If a data-sequence contains some sequence, it contains every contiguous subsequence of that sequence. Thus, if a data-sequence contains some sequence of length greater than k, it contains at least one sequence of length $k + 1$.

Theorem 5. *Sequential patterns, to be returned by a data mining query, containing a given sequence can be contained only in data-sequences containing that sequence using min-gap and window-size specified in the query, and max-gap of $+\infty$.*

Proof. If a data-sequence contains some sequence using certain values of max-gap, min-gap, and window-size, it also contains every contiguous subsequence of the sequence, using the same time constraints. If max-gap is set to $+\infty$, a data-sequence containing some sequence contains all its subsequences.

Theorem 6. *Sequential patterns, to be returned by a data mining query, whose n-th element has the size greater than k can be contained only in data-sequences which contain some 1-element sequence of size $k + 1$ using window-size specified in the query.*

Proof. Each 1-element subsequence of any sequence is its contiguous subsequence (from the definition of a contiguous subsequence). If any element of a sequence has the size greater than k, the sequence has at least one 1-element contiguous subsequence of size $k + 1$. If a data-sequence contains some sequence, it contains every contiguous subsequence of that sequence. Thus, if a data-sequence contains some sequence whose n-th element has the size greater than k, it has to contain some 1-element sequence of size $k + 1$.

Theorem 7. *Sequential patterns, to be returned by a data mining query, whose n-th element contains a given set can be contained only in data-sequences which contain a 1-element sequence having the set as the only element, using time constraints specified in the query.*

Proof. Each 1-element subsequence of any sequence is its contiguous subsequence (from the definition of a contiguous subsequence). If any element of a sequence contains a given set, a 1-element sequence formed by the set is a contiguous subsequence of the sequence. If a data-sequence contains some sequence, it contains every contiguous subsequence of that sequence. Thus, if a data-sequence contains some sequence whose n-th element contains a given set, it has to contain a 1-element sequence having the set as the only element.

According to the above theorems the following dataset filtering predicates are applicable in sequential pattern mining:

- $\sigma(\mathbf{SG}, \alpha, sequence)$ - true if the size of the data-sequence is greater than α, false otherwise;
- $\sigma(\mathbf{C}, \beta, sequence, maxgap, mingap, window)$ - true if the data-sequence contains the sequence forming the pattern β using given time constraints, false otherwise;
- $\sigma(\mathbf{CS}, \alpha, sequence, window)$ - true if there exists a 1-element sequence of size α that is contained in the sequence with respect to the window-size constraint, false otherwise;
- $\sigma(\mathbf{CL}, \alpha, sequence, maxgap, mingap, window)$ - true if there exists a sequence of length α that is contained in the sequence with respect to the max-gap, min-gap, and window-size constraints, false otherwise.

For a sequential pattern query with pattern constraints, an appropriate dataset filtering predicate is derived in the following way: For each of the pattern predicates from the left column of Table 3 present in the query, the corresponding data-sequence predicate from the right table column is added to the dataset filtering predicate.

In the above table, $< \gamma >$ denotes a 1-element sequence having the set γ as its only element, while max, min, and win represent values of max-gap, min-gap, and window-size time constraints respectively.

Table 3. Derivation rules for sequential pattern mining

Pattern predicate	Data-sequence predicate
$\pi(\mathbf{SG},\ \alpha,\ pattern)$	$\sigma(\mathbf{SG},\ \alpha,\ sequence)$
$\pi(\mathbf{LG},\ \alpha,\ pattern)$	$\sigma(\mathbf{CL},\ \alpha + 1,\ sequence,\ max,\ min,\ win)$
$\pi(\mathbf{C},\ \beta,\ pattern)$	$\sigma(\mathbf{C},\ \beta,\ sequence,\ +\infty,\ min,\ win)$
$\rho(\mathbf{SG},\ \alpha,\ pattern_n)$	$\sigma(\mathbf{CS},\ \alpha + 1,\ sequence,\ win)$
$\rho(\mathbf{C},\ \gamma,\ pattern_n)$	$\sigma(\mathbf{C},\ <\gamma>,\ sequence,\ max,\ min,\ win)$

4   Implementation Issues Regarding Dataset Filtering

If any pattern predicates supporting dataset filtering are present in the pattern query specifying a certain pattern discovery task, the query can be transformed into a query representing a discovery task on a potentially smaller dataset in the following way. Firstly, database constraints of the query have to be extended by adding the appropriate dataset filtering predicate to them (the filtering predicate is derived according to the rules presented in the previous section). Secondly, the minimum support threshold has to be adjusted to the size of the filtered database. This step is necessary because the support of a pattern is expressed as the percentage of objects (transactions or data-sequences) containing the pattern. The theorems proved in the previous section guarantee that the number of objects containing a given pattern in the original and filtered dataset will be the same as long as the pattern satisfies pattern constraints. Thus, we have the following relationship between the support of a pattern p (satisfying pattern constraints) in the original and filtered datasets: $sup_F(p) = |D| * sup(p)/|D_F|$, where $sup_F(p)$ and $sup(p)$ denote the support of the pattern p in the filtered and original dataset respectively, and $|D_F|$ and $|D|$ denote the number of objects in the filtered and original dataset respectively. After the patterns frequent in the filtered dataset have been discovered, their support has to be normalized with respect to the number of objects in the original dataset according to the above formula (the user specifies the support threshold as the percentage of objects in the original dataset, and expects that the supports of discovered patterns will be expressed in the same way).

Dataset filtering techniques can be combined with any frequent pattern discovery algorithm since they conceptually lead to a transformed discovery task guaranteed to return the same set of patterns as the original task. The transformation of the source dataset (by filtering out objects that cannot contain patterns of interest) and adjustment of the minimum support threshold can be performed before the actual discovery process. However, in reality such explicit transformation might be impossible due to space limitations. Moreover, it may not lead to the optimal solution because of one extra scan of the dataset performed during the transformation. A natural solution to this problem is integration of dataset filtering techniques within pattern mining algorithms. Thus, dataset filtering can be performed together with other operations in the first scan of the dataset required by a given algorithm. Please note that if dataset filtering is integrated within a pattern mining algorithm, the support conversions discussed above are not necessary because the support can always refer to the number of objects in the original dataset.

Regarding integration of dataset filtering within the *Apriori* framework, there are two general implementation strategies possible. The filtered dataset can either be physically materialized on disk during the first iteration or filtering can be performed on-line in each iteration. The second option might be the only solution if materialization of the filtered dataset is not possible due to space limitations.

Below we present two algorithm frameworks following the two strategies. We do not present separate solutions for frequent itemset and sequential pattern discovery. Instead, we consider a general pattern discovery task in a collection of objects. These objects are transactions or data-sequences depending on the actual discovery task. Pattern constraints and dataset filtering predicates are also different for frequent itemsets and sequential patterns. Additionally, in the context of sequential patterns the containment relationship takes into account time constraints specified by a user. The general algorithms presented below take a collection D of objects, the minimum support threshold (and optionally time constraints), and pattern constraints as input, and return all frequent patterns in D satisfying all the provided constraints.

Algorithm 1 Apriori on materialized filtered dataset
begin
    DF = dataset filtering predicate derived from pattern constraints;
    scan D in order to:
        1) evaluate minimum number of supporting
        objects for a pattern to be called frequent (*mincount*)
        2) find L_1 (set of items contained in at
        least *mincount* objects satisfying DF);
        3) materialize the collection D' of objects from D
        satisfying DF;
    for $(k = 2; L_{k-1} \neq \emptyset; k{+}{+})$ **do**
    begin
        C_k = apriori_gen(L_{k-1}); /* generate new candidates */
        if $C_k = \emptyset$ **then break**;
        forall objects $d \in D'$ **do**
            forall candidates $c \in C_k$ **do**
                if d contains c **then**
                      c.count ++;
                end if;
        $L_k = \{\, c \in C_k \mid c\text{.count} \geq mincount\}$;
    end;
    output patterns from $\cup_k L_k$ satisfying pattern constraints;
end.

Algorithm 2 Apriori with on-line dataset filtering

begin

    DF = dataset filtering predicate derived from pattern constraints;

    scan D in order to:

        1) evaluate minimum number of supporting

        objects for a pattern to be called frequent (*mincount*)

        2) find L_1 (set of items contained in at

        least *mincount* objects satisfying DF);

    for $(k = 2; L_{k-1} \neq \emptyset; k{+}{+})$ **do**

    begin

        C_k = apriori_gen(L_{k-1}); /* generate new candidates */

        if $C_k = \emptyset$ **then break**;

        forall objects $d \in D$ **do**

            if d satisfies DF **then**

                forall candidates $c \in C_k$ **do**

                    if d contains c **then**

                        c.count $++$;

                end if;

            end if;

        L_k = { $c \in C_k \mid c$.count $\geq$ *mincount*};

    end;

    output patterns from $\cup_k L_k$ satisfying pattern constraints;

end.

Both algorithms start with deriving dataset filtering predicates from pattern constraints provided by a user. In the first approach these dataset filtering predicates are used in the first scan of the source dataset to select and materialize the collection of objects on which subsequent scans will be performed. All the objects from the materialized filtered collection are then used in the candidate verification phases. In the second approach dataset filtering predicates are used in each scan of the source dataset and objects that do not satisfy them are excluded from the candidate verification process. When the discovery of sequential patterns in the filtered dataset is finished, a post-processing step filtering out patterns that do not satisfy user-specified pattern constraints is applied in both approaches. This phase is required since dataset filtering itself, regardless of the implementation details, does not guarantee that only patterns supporting pattern constraints are to be discovered. It should be noted that the support of patterns not satisfying user-specified pattern constraints, counted in the filtered dataset, can be smaller than their actual support in the original dataset, but it is not a problem since these patterns will not be returned to the user. Moreover, this is in fact a positive feature as it can reduce the number of generated candidates not leading to patterns of user's interest.

5   Performance Analysis

In order to evaluate performance gains offered by our dataset filtering techniques, we performed several experiments on synthetic datasets. We measured performance improvements thanks to dataset filtering applied to the *Apriori* algorithm for frequent itemset mining, and *GSP* for sequential patterns. As we might have expected, the performance gains depend on the selectivity of dataset filtering predicates derived from pattern constraints (expressed as the percentage of objects in the dataset satisfying dataset filtering constraints). In general the lower the selectivity factor the better, but the actual performance depends not only on the selectivity but also on data distribution in the filtered dataset. We observed that item constraints led to much better results (reducing the processing time 2 to 5 times) than size constraints (typically reducing the processing time by less than 10%). This is due to the fact that the patterns are usually smaller in terms of size or length than source objects, and therefore even restrictive constraints on pattern size/length result in weak constraints on source objects. As a consequence, if the actual task is discovery of association rules, and the only rule constraints present are size constraints, the gains due to dataset filtering sometimes do not compensate the cost of an extra pass needed to evaluate confidences of the rules.

Regarding the implementation strategies, for item constraints implementations involving materialization of the filtered dataset were more efficient than their on-line counterparts (the filtered dataset was relatively small and the materialization cost was dominated by gains due to the smaller costs of dataset scans in candidate verification phases). However, in case of size constraints rejecting a very small number of source objects, materialization of the filtered dataset sometimes lead to longer execution times than in case of the original algorithms. The on-line dataset filtering implementations were in general more efficient than the original algorithms even for size constraints (except for a situation, unlikely in practice, when the size constraint did not reject any source objects).

In the experiments, we also observed that decreasing the minimum support threshold or relaxing time constraints worked in favor of our dataset filtering techniques, leading to bigger performance gains. This behavior can be explained by the fact that since dataset filtering reduces the cost of candidate verification phase, the more this phase contributes to the overall processing time, the more significant relative performance gains are going to be. Decreasing the minimum support threshold also led to slight performance improvement of implementations involving materialization of the filtered dataset in comparison to their on-line counterparts. As the support threshold decreases, the maximal length of a frequent patterns (and the number of iterations required by the algorithms) increases. Materialization is performed in the first iteration and reduces the cost of the second and subsequent iterations. Thus, the more iterations are required, the better the cost of materialization is compensated.

6 Concluding Remarks

We have discussed application of dataset filtering techniques to efficient frequent pattern mining in the presence of various pattern constraints. We identified the types of pattern constraints in case of which dataset filtering is applicable in the context of frequent itemsets, association rules, and sequential patterns. For each of the pattern constraint types we provided an appropriate dataset filtering predicate. Dataset filtering techniques can be applied to any frequent pattern discovery algorithm since they conceptually lead to an equivalent data mining task on a possibly smaller dataset. We focused on the implementation details concerning integration of dataset filtering techniques within the *Apriori* framework. Our experiments show that dataset filtering can result in significant performance improvements, especially in case of pattern constrains involving the presence of a certain subset or subsequence, which we believe are the most usful ones.

References

1. Agrawal R., Imielinski T., Swami A.: Mining Association Rules Between Sets of Items in Large Databases. Proc. of the 1993 SIGMOD Conference (1993)
2. Agrawal R., Srikant R.: Fast Algorithms for Mining Association Rules. Proc. of the 20th VLDB Conference (1994)
3. Agrawal R., Srikant R.: Mining Sequential Patterns. Proc. of the 11th ICDE Conf. (1995)
4. Garofalakis M., Rastogi R., Shim K.: SPIRIT: Sequential Pattern Mining with Regular Expression Constraints. Proceedings of 25th VLDB Conference (1999)
5. Han J., Lakshmanan L., Ng R.: Constraint-Based Multidimensional Data Mining. IEEE Computer, Vol. 32, No. 8 (1999)
6. Han J., Pei J.: Mining Frequent Patterns by Pattern-Growth: Methodology and Implications. SIGKDD Explorations, December 2000 (2000)
7. Imielinski T., Mannila H.: A Database Perspective on Knowledge Discovery. Communications of the ACM, Vol. 39, No. 11 (1996)
8. Ng R., Lakshmanan L., Han J., Pang A.: Exploratory Mining and Pruning Optimizations of Constrained Association Rules. Proc. of the 1998 SIGMOD Conference (1998)
9. Pei J., Han J., Lakshmanan L.: Mining Frequent Itemsets with Convertible Constraints. Proceedings of the 17th ICDE Conference (2001)
10. Srikant R., Agrawal R.: Mining Sequential Patterns: Generalizations and Performance Improvements. Proc. of the 5th EDBT Conference (1996)
11. Srikant R., Vu Q., Agrawal R.: Mining Association Rules with Item Constraints. Proceedings of the 3rd KDD Conference (1997)
12. Zheng Z., Kohavi R., Mason L.: Real World Performance of Association Rule Algorithms. Proc. of the 7th KDD Conference (2001)

Concise Representations of Association Rules

Marzena Kryszkiewicz

Institute of Computer Science, Warsaw University of Technology
Nowowiejska 15/19, 00-665 Warsaw, Poland
mkr@ii.pw.edu.pl

Abstract. Strong association rules are one of basic types of knowledge. The number of rules is often huge, which limits their usefulness. Applying concise rule representations with appropriate inference mechanisms can lessen the problem. Ideally, a rule representation should be lossless (should enable derivation of all strong rules), sound (should forbid derivation of rules that are not strong) and informative (should allow determination of rules' support and confidence). In the paper, we overview the following lossless representations: representative rules, Duquenne-Guigues basis, proper basis, Luxemburger basis, structural basis, minimal non-redundant rules, generic basis, informative basis and its transitive reduction. For each representation, we examine whether it is sound and informative. For the representations that are not sound, we discuss ways of turning them into sound ones. Some important theoretical results related to the relationships among the representations are offered as well.

1 Introduction

Strong association rules are one of basic types of knowledge discovered from large databases. The number of association rules is often huge, which limits their usefulness in real life applications. A possible solution to this problem is to restrict extraction of rules to those that are strictly related to user's needs. Statistical measures can be used to determine whether a rule is useful for predictive purposes or reveals surprising relationships. Such measures are usually heuristic and subjective. Another approach to the problem, we will consider in this paper, consists in applying appropriate inference mechanisms to concise representations of strong association rules. In an ideal case, a rule representation should be *lossless* (should enable derivation of all strong rules), *sound* (should forbid derivation of rules that are not strong) and *informative* (should allow determination of rules parameters such as support and confidence). Let us note that the usefulness of a lossless rule representation that admits identification of non-strong rules as strong ones is questionable. In the result, the user provided with such a set of rules will not know which of them are correct and which are not, although the set contains all strong rules. The aspect of soundness of rule representations has not been captured except for the representative rules [4,7]. In the paper, we will overview the following lossless representations of strong association rules:

D.J. Hand et al. (Eds.): Pattern Detection and Discovery, LNAI 2447, pp. 92–109, 2002.

representative rules[1], Duquenne-Guigues basis, proper basis, Luxenburger basis, structural basis [10,11], minimal non-redundant rules, generic basis, informative basis and its transitive reduction [3,10].[2] For each representation, we examine whether it is sound and informative. For the representations that are not sound, we discuss ways of turning (extending) them into sound ones. Some important theoretical results related to the relationships among the representations are offered as well.

2 Basic Notions: Association Rules, Frequent Itemsets, Closed Itemsets, and Generators

The notion of association rules and the problem of their discovering were introduced in [2] for sales transaction database. Association rules identify sets of items that are purchased together with other sets of items. For example, an association rule may state that 80% of customers buy fish also buy white wine.

Definition 2.1. Let $I = \{i_1, i_2, \ldots, i_m\}$ be a set of distinct literals, called *items* (e.g. items to be sold). Any subset X of items in I is called an *itemset*; that is $X \subseteq I$. A *transaction database*, denoted by $\mathcal{D}$, is a set of itemsets. Each itemset T in $\mathcal{D}$ is a *transaction*. An *association rule* is an expression associating two itemsets:

$$X \to Y, \text{ where } \emptyset \neq Y \subset I \text{ and } X \subseteq I \backslash Y.$$

X is called the *antecedent*, and Y is called the *consequent* of the rule. The rule $X \to Y$ is said to be *based on* $X \cup Y$, and $X \cup Y$ is called the *base* of $X \to Y$. $\mathcal{AR}$ will denote the set of all association rules that can be created from itemsets $X \subseteq I$. $\diamond$

Itemsets and association rules are characterized by simple statistical parameters stating their importance or strength.

Definition 2.2. Statistical significance of an itemset X is called *support* and is denoted by $sup(X)$. $sup(X)$ is defined as the number of transactions in $\mathcal{D}$ that contain X; that is $sup(X) = |\{T \in \mathcal{D} | X \subseteq \mathcal{D}\}|$. Statistical significance (*support*) of a rule $X \to Y$ is denoted by $sup(X \to Y)$ and is defined as support of the base $X \cup Y$ of the rule $X \to Y$, i.e. $sup(X \to Y) = sup(X \cup Y)$. The *confidence* (strength) of an association rule $X \to Y$ is denoted by $conf(X \to Y)$ and is defined as the conditional probability that Y occurs in a transaction provided X occurs in the transaction; that is $conf(X \to Y) = sup(X \to Y)/sup(X)$. $\diamond$

Usually, one is interested in discovering patterns and association rules of high support and confidence.

---

[1] Independently in 1998, it was proposed in [1] to discard so called simple redundant and strict redundant association rules, which strictly correspond to the association rules that are not representative. In the result, the set of remaining association rules is equal to the set of representative rules as defined in [4]. Representative rules were also rediscovered in 2001 in [8] under the name of *representative basis*.

[2] In our review we skipped the representation from [12] as it is currently in an early stage - subject to further research.

Definition 2.3. Itemsets whose support is above the minimum support threshold $minSup$ are denoted by $\mathcal{F}$ and called *frequent*; i.e. $\mathcal{F} = \{X \subseteq I | sup(X) > minSup\}$. *Strong association rules* are denoted by $\mathcal{AR}$ and are defined as association rules whose support is above the minimum support threshold $minSup$ ($minSup \in [0, |\mathcal{D}|)$) and confidence is above the minimum support confidence $minConf$ ($minConf \in [0, 1)$); that is:

$$\mathcal{AR} = \{r \in AR | sup(r) > minSup \wedge conf(r) > minConf\}. \qquad \diamond$$

The minimum support thresholds $minSup$ and $minConf$ are supposed to be defined by a user. The problem of deriving strong association rules from the database is usually decomposed into two subproblems:

1. Generate all frequent itemsets $\mathcal{F}$.
2. Generate all association rules from $\mathcal{F}$. Let $Z \in \mathcal{F}$ and $\emptyset \neq X \subset Z$. Then any candidate rule $X \Rightarrow Z \setminus X$ is association one if $sup(Z)/sup(X) > minConf$.

Efficient solutions to these subproblems apply the following important property of itemsets, which reduces evaluations of candidate itemsets and candidate rules considerably: All subsets of a frequent itemset are frequent and all supersets of an infrequent itemset are infrequent. The property follows the observation that $sup(X) \geq sup(Y)$ for $X \subset Y$.

A number of concise lossless representations have been proposed for both frequent itemsets (see e.g. [6,7] for an overview) and strong association rules. The representations based on *generators* and *closed itemsets*, which, although not the most concise ones, play an important role among the representations of frequent itemsets, since they can be applied for direct derivation of the majority of association rules representations. For example, in the case of representative association rules [5] and informative basis [3], the antecedent of any such rule is a generator, while the consequent is a closed itemset decreased by the items present in the rule's antecedent. Below we recollect the notions of closed itemsets and generators:

Definition 2.4. Let $X \subseteq I$. We define the *closure operator* $\gamma(X)$ as the intersection of transactions in $\mathcal{D}$ containing X, i.e.: $\gamma(X) = \bigcap\{T \in \mathcal{D} | T \supseteq X\}$. An itemset X is *closed* if $\gamma(X) = X$. The set of all closed itemsets is denoted by $\mathcal{C}$. Let X be a closed itemset. A minimal itemset Y satisfying $\gamma(Y) = X$ is called a generator of X. By $\mathcal{G}(X)$ we will denote *the set of all generators of* X. The union of generators of all closed itemsets will be denoted by $\mathcal{G}$, i.e. $\mathcal{G} = \bigcup_{X \in \mathcal{C}} \mathcal{G}(X)$. $\diamond$

Clearly, $\gamma(X)$ is the greatest (w.r.t. set inclusion) itemset that occurs in all transactions in $\mathcal{D}$ in which X occurs and has the same support as X, while each generator in $\mathcal{G}(X)$ is such a minimal itemset.

Property 2.1 [7]. Let $X \subseteq I$.
a) If $X \subset Y$ and $sup(Y) = sup(X)$, then $\gamma(Y) = \gamma(X)$,
b) $X \in \mathcal{C}$ iff $\forall Y \subseteq I$, if $Y \supset X$, then $sup(X) \neq sup(Y)$,
c) $X \in \mathcal{G}$ iff $\forall Y \subseteq I$, if $Y \subset X$, then $sup(X) \neq sup(Y)$,
d) $X \subseteq Z$ iff $\gamma(X) \subseteq Z$ for all $Z \in \mathcal{C}$.

Definition 2.5. *Frequent closed itemsets* will be denoted by FC, that is $FC = \mathcal{F} \cap \mathcal{C}$. Frequent generators will be denoted by FG, that is $FG = \mathcal{F} \cap \mathcal{G}$. $\diamond$

3   Rule Inference Mechanisms

This section presents *rule inference mechanisms* understood as methods of deriving association rules based on other rules. We start with reminding *Armstrong's axioms* and the rule *confidence transitivity property* [9]. Then we will present the notion of a cover operator we introduced in [4]. Finally, we present a method of reasoning with rules based on closed itemsets that was addressed in [3,10,11, 12].

3.1   Armstrong's Axioms (AA)

Armstrong Axioms relate only to association rules that are certain:

Property 3.1.1 (of Armstrong's axioms).
a) $conf(X \to X) = 1$
b) $conf(X \to Y) = 1 \Rightarrow conf(X \cup Z \to Y) = 1,$
c) $conf(X \to Y) = 1 \wedge conf(Y \cup Z \to W) = 1 \Rightarrow conf(X \cup Z \to W) = 1.$

3.2   Confidence Transitivity Property (CTP)

The property of rules inference was proposed by Luxenburger [9]. Unlike Armstrong axioms, it can be applied to association rules that are not certain.

Property 3.2.1 (of confidence transitivity). Let $X \subset Y \subset Z$. Support and confidence of the rule $X \to Z \backslash X$ can be derived from confidences of the rules: $X \to Y \backslash X$ and $Y \to Z \backslash Y$, as follows:
a) $sup(X \to Z \backslash X) = sup(Y \to Z \backslash Y),$
b) $conf(X \to Z \backslash X) = conf(X \to Y \backslash X) \times conf(Y \to Z \backslash Y),$

The confidence of the rule derived by applying confidence transitivity inference is never greater than the confidences of rules from which it was calculated.

3.3   Cover Operator (C)

In [4] we introduced the concept of cover operator, which given a strong association rule $X \to Y$, derives a set of strong association rules by simple syntactical transforming of $X \to Y$.

Definition 3.3.1. The *cover* C of an association rule $X \to Y$, is defined as:

$$C(X \to Y) = \{X \cup Z \to V | Z, V \subseteq Y \wedge Z \cap V = \emptyset \wedge V \neq \emptyset\}. \qquad \diamond$$

Each rule in $C(X \to Y)$ consists of a subset of items occurring in the rule $X \to Y$. The antecedent of any rule r covered by $X \to Y$ contains X and perhaps some items in Y, whereas $r's$ consequent is a non-empty subset of the remaining items in Y. In other words, $C(X \to Y)$ contains all rules obtained from $X \to Y$ by deleting a subset of items from Y (i.e. strict redundant rules in terminology from [1]) and/or moving a subset of items from Y to X (i.e. simple redundant rules in terminology from [1]). For the cover C the following property holds:

Proposition 3.3.1 (of cover inference) [4]. Let r be an association rule.

a) If $r' \in C(r)$, then $sup(r') \geq sup(r)$ and $conf(r') \geq conf(r)$,

b) If $r \in \mathcal{AR}$ and $r' \in C(r)$, then $r' \in \mathcal{AR}$,

c) If $r \in \mathcal{AR}$, then $C(r) \subseteq \mathcal{AR}$.

So, C derives only strong rules from each strong association rule.

Table 1. Example database $\mathcal{D}$

Id	Transaction
T_1	$\{abcd\}$
T_2	$\{abcdef\}$
T_3	$\{abcdehi\}$
T_4	$\{abe\}$
T_5	$\{bcdehi\}$

Table 2. The cover of the rule $r : (\{b\} \to \{de\})$

#	Rule r' in $C(r)$	Support of r'	Confidence of r'
1.	$\{b\} \to \{de\}$	4	80%
2.	$\{b\} \to \{d\}$	4	80%
3.	$\{b\} \to \{e\}$	5	100%
4.	$\{bd\} \to \{e\}$	4	100%
5.	$\{be\} \to \{d\}$	4	80%

Example 3.3.1. Let us consider the database from Table 1. Let $r : \{b\} \to \{de\}$. Then, $C(r) = \{\{b\} \to \{de\}, \{b\} \to \{d\}, \{b\} \to \{e\}, \{bd\} \to \{e\}, \{be\} \to \{d\}\}$ (see also Table 2). The support of r is equal to 4 and its confidence is equal to 80%. The support and confidence of all other rules in $C(r)$ are not less than the support and confidence of r. □

The next proposition states that the number of association rules covered by a given rule r depends only on the number of items in the consequent of r.

Proposition 3.3.2 [4]. $|C(X \to Y)| = 3^m - 2^m$, where $m = |Y|$.

Example 3.3.2. Let us consider association rule $r : (\{b\} \to \{de\})$ (see Table 2). Then, $|C(r)| = 3^2 - 2^2 = 9 - 4 = 5$. Thus, r represents 5 association rules. □

Proposition 3.3.3 [4]. Let $r : (X \to Y)$ and $r' : (X' \to Y')$ be association rules. The following statements are equivalent:

a) $r' \in C(r)$,

b) $X' \cup Y' \subseteq X \cup Y \wedge X' \supseteq X$,

c) $X' \cup Y' \subseteq X \cup Y \wedge X' \supseteq X \wedge Y' \subseteq Y$.

Corollary 3.3.1. Let $r \neq r'$. If $r \in C(r')$, then $r' \notin C(r)$.

3.4   Closure-Closure Rule Inference (CCI)

Closure-closure rule inference is based on the observation that the support of an itemset is equal to the support of its closure:

Property 3.4.1 (of closure-closure rule inference). Let $X \to Y \backslash X$, where $X \subset Y$, be a rule under consideration. Then:

a) $sup(X \to Y \backslash X) = sup(\gamma(Y))$,

b) $conf(X \to Y \backslash X) = sup(\gamma(Y)) / sup(\gamma(X))$,

c) $(X \to Y \backslash X) \in \mathcal{AR}$ if $(\gamma(X) \to \gamma(Y) \backslash \gamma(X)) \in \mathcal{AR}$.

Property 3.4.2 (of closure determination) [3]. Let X be an itemset. The closure $\gamma(X)$ is equal to the smallest (w.r.t. set inclusion) closed itemset containing X.

4   Classes of Representations of Association Rules

Definition 4.1. A pair $(R, \models)$, where $R \subseteq AR$ and $\models$ is a rule inference mechanism, is called a *representation of association rules*.   ◇

Let AR' be a subset (proper or improper) of association rules. We define the following classes of representations of association rules.

Definition 4.2. $(R, \models)$ is called *lossless representation of AR'* if $\models$ derives all association rules in AR' from R. $(R, \models)$ is called *sound representation of AR'* if all rules derivable from R by $\models$ belong to AR'. $(R, \models)$ is called *informative representation* if $\models$ determines support and confidence for each rule derivable from R.   ◇

In general, the proposed rule classes are not mutually exclusive. In fact, it is desirable so that a representation $(R, \models)$ of strong association rules AR belongs to all three categories. However, if the representation $(R, \models)$ is not informative, but is lossless and sound, then it is still useful in that it provides all and only strong association rules. It means it derives all and only rules required by the user. The representation that is sound is also of value as it derives some, though not all, association rules. The representation $(R, \models)$ that is lossless and informative, but not sound, can be easily transformed into a lossless, informative, and sound one. It is enough to discard those derived rules that have support or confidence below the required threshold. On the other hand, the representation $(R, \models)$ of AR, which is lossless, but is neither sound nor informative, is of no value. Although such a representation derives a superset of strong association rules, there is no method to determine, which of the derived rules are strong and which are not. Hence, the user does not know, which rules are sufficiently "true" and provide important knowledge.

5   Lossless Representations of Association Rules

In [4], we introduced a concept of *representative association rules* as a lossless and sound representation of strong association rules. The other lossless representations of strong association rules that were proposed in the literature distinguish between *certain (exact)* association rules, whose confidences are equal to 1, and *approximate* rules, whose confidences are less than 1. The corresponding representations consist of two parts: one that represents all certain rules, and another one that represents all approximate rules. In the case of two-part representations, soundness have not been considered so far.

5.1   Representative Association Rules

Having C as a mechanism for inferring rules of not worse quality than a given rule, it is justified to restrict generation of AR to a set of rules covering all others. In the sequel, such a minimal base of rules will be called *representative association rules* (or briefly *representative rules*) and will be denoted by RR.

Definition 5.1.1. *Representative rules* $\mathcal{RR}$ are those strong association rules that are not covered by other strong association rules:

$$\mathcal{RR} = \{r \in \mathcal{AR} | \neg \exists r' \in \mathcal{AR}, r' \neq r \wedge r \in C(r')\}. \qquad \diamond$$

By definition of $\mathcal{RR}$, no representative association rule belongs to the cover of another strong association rule. On the other hand, $\mathcal{RR}$s are sufficient to derive all strong association rules by applying the cover operator:

Lemma 5.1.1 [7]. $\forall r \in \mathcal{AR} \, \exists r' \in \mathcal{RR}$ such that $r \in C(r')$.

Theorem 5.1.1 [7]. $\mathcal{AR} = \bigcup_{r \in \mathcal{RR}} C(r)$.

Proof. By Lemma 5.1.1, $\bigcup_{r \in \mathcal{RR}} C(r) \supseteq \mathcal{AR}$ (∗). By Proposition 3.3.1, $\forall r \in \mathcal{RR} \; C(r) \subseteq \mathcal{AR}$ (∗∗). Hence, by (∗) and (∗∗), $\bigcup_{r \in \mathcal{RR}} C(r) = \mathcal{AR}$. $\qquad \square$

Corollary 5.1.1. $(\mathcal{RR}, C)$ is a lossless and sound representation of $\mathcal{AR}$s.

In general, the pair $(\mathcal{RR}, C)$ is not informative. However, if a rule r belongs to the cover of several representative rules, say $\mathcal{R} \subseteq \mathcal{RR}$, then $sup(r) \geq max\{sup(r) | r \in C(r'), r' \in \mathcal{R}\}$, and $conf(r) \geq max\{conf(r) | r \in C(r'), r' \in \mathcal{R}\}$. This observation may improve estimation of support and confidence of rules that are not representative.

Observe that in order to determine whether $X \rightarrow Z \backslash X$ is strong or not, it is not necessary to calculate covers of rules in $\mathcal{RR}$ explicitly. According to Lemma 5.1.1 and Propositions 3.3.1, 3.3.3, $X \rightarrow Z \backslash X$ is strong if there is $X' \rightarrow Z' \backslash X'$ in $\mathcal{RR}$ such that $X' \subseteq X$ and $Z' \supseteq Z$. Hence, determination if $X \rightarrow Z \backslash X$ is strong, can be carried out by matching a candidate rule against representative rules, and is linear in the worst case.

Example 5.1.1. Given $minSup = 2$ and $minConf = 77\%$, the following representative rules would be found for the database $\mathcal{D}$ from Table 1 (see also Table 3): $\mathcal{RR} = \{\emptyset \rightarrow \{abe\}[4, 4/5], \emptyset \rightarrow \{bcde\}[4, 4/5], \{ac\} \rightarrow \{bde\}[3, 1], \{ad\} \rightarrow \{bce\}[3, 1]\}$. There are 4 representative association rules in $\mathcal{RR}$, whereas the number of all association rules in $\mathcal{AR}$ is 112. Hence, $\mathcal{RR}$ constitutes 3.57% of $\mathcal{AR}$. Let us now test if the rules $\{b\} \rightarrow \{e\}$ and $\{a\} \rightarrow \{bc\}$ are strong. The rule $\{b\} \rightarrow \{e\}$ is strong because it belongs to the cover of, for example, the representative rule $\emptyset \rightarrow \{bcde\}$. The second rule $\{a\} \rightarrow \{bc\}$ is not strong, as there is no representative rule covering it, i.e. having a subset of $\{a\}$ as an antecedent and a superset of $\{abc\}$ as a base. $\qquad \square$

The proposition below states that each representative rule is based on a frequent closed itemset and its antecedent is a frequent generator.

Proposition 5.1.1 [5,7]. If $(X \rightarrow Z \backslash X) \in \mathcal{RR}$, then $X \in FG$ and $Z \in FC$.

5.2   Generic Basis, Informative Basis, and Related Representations

In this subsection, we recall after [3,10] a concept of minimal non-redundant association rules and related representations of association rules. Next we examine these concepts more thoroughly showing their strong and weak points and proposing solutions to the weak points.

Table 3. The representations of strong association rules mined in the database from Table 1 given $minSup = 2$ and $minConf = 77\%$, 70% and 50%, respectively

Representation	$minConf = 77\%$	$minConf = 70\%$	$minConf = 50\%$
$\mathcal{RR}$	$\emptyset \to \{abe\}[4, 4/5]$	$\emptyset \to \{abe\}[4, 4/5]$	$\emptyset \to \{abcde\}[3, 3/5]$
	$\emptyset \to \{bcde\}[4, 4/5]$	$\emptyset \to \{bcde\}[4, 4/5]$	
	$\{ac\} \to \{bde\}[3, 1]$	$\{a\} \to \{bcde\}[3, 3/4]$	
	$\{ad\} \to \{bce\}[3, 1]$	$\{c\} \to \{abde\}[3, 3/4]$	
		$\{d\} \to \{abce\}[3, 3/4$	
$\mathcal{GB}$	$\emptyset \to \{be\}[5, 1]$	$\emptyset \to \{be\}[5, 1]$	$\emptyset \to \{be\}[5, 1]$
	$\{a\} \to \{be\}[4, 1]$	$\{a\} \to \{be\}[4, 1]$	$\{a\} \to \{be\}[4, 1]$
	$\{c\} \to \{bde\}[4, 1]$	$\{c\} \to \{bde\}[4, 1]$	$\{c\} \to \{bde\}[4, 1]$
	$\{d\} \to \{bce\}[4, 1]$	$\{d\} \to \{bce\}[4, 1]$	$\{d\} \to \{bce\}[4, 1]$
	$\{ac\} \to \{bde\}[3, 1]$	$\{ac\} \to \{bde\}[3, 1]$	$\{ac\} \to \{bde\}[3, 1]$
	$\{ad\} \to \{bce\}[3, 1]$	$\{ad\} \to \{bce\}[3, 1]$	$\{ad\} \to \{bce\}[3, 1]$
$\mathcal{IB}$	$\emptyset \to \{abe\}[4, 4/5]$	$\emptyset \to \{abe\}[4, 4/5]$	$\emptyset \to \{abe\}[4, 4/5]$
	$\emptyset \to \{bcde\}[4, 4/5]$	$\emptyset \to \{bcde\}[4, 4/5]$	$\emptyset \to \{bcde\}[4, 4/5]$
		$\{a\} \to \{bcde\}[3, 3/4]$	$\{c\} \to \{bcde\}[3, 3/4]$
		$\{c\} \to \{abde\}[3, 3/4]$	$\{a\} \to \{abde\}[3, 3/4]$
		$\{d\} \to \{abce\}[3, 3/4]$	$\{d\} \to \{abce\}[3, 3/4]$
$\mathcal{RI}$	$\emptyset \to \{abe\}[4, 4/5]$	$\emptyset \to \{abe\}[4, 4/5]$	$\emptyset \to \{abe\}[4, 4/5]$
	$\emptyset \to \{dcde\}[4, 4/5]$	$\emptyset \to \{bcde\}[4, 4/5]$	$\emptyset \to \{bcde\}[4, 4/5]$
		$\{a\} \to \{bcde\}[3, 3/4]$	$\{a\} \to \{bcde\}[3, 3/4]$
		$\{a\} \to \{abde\}[3, 3/4]$	
		$\{a\} \to \{abce\}[3, 3/4]$	
$\mathcal{DG}$	$\emptyset \to \{be\}[5, 1]$	$\emptyset \to \{be\}[5, 1]$	$\emptyset \to \{be\}[5, 1]$
	$\{bce\} \to \{d\}[4, 1]$	$\{bce\} \to \{d\}[4, 1]$	$\{bce\} \to \{d\}[4, 1]$
	$\{bde\} \to \{c\}[4, 1]$	$\{bde\} \to \{c\}[4, 1]$	$\{bde\} \to \{c\}[4, 1]$
$\mathcal{PB}$ (or $\mathcal{LB}$)	$\emptyset \to \{abe\}[4, 4/5$	$\emptyset \to \{abe\}[4, 4/5$	$\emptyset \to \{abe\}[4, 4/5$
	$\emptyset \to \{bcde\}[4, 4/5$	$\emptyset \to \{bcde\}[4, 4/5$	$\emptyset \to \{bcde\}[4, 4/5$
		$\{abe\} \to \{cd\}[3, 3/4]$	$\{abe\} \to \{cd\}[3, 3/4]$
		$\{bcde\} \to \{a\}[3, 3/4]$	$\{bcd\} \to \{a\}[3, 3/4]$
			$\{be\} \to \{acd\}[3, 3/5]$
$\mathcal{SB}$	$\emptyset \to \{abe\}[4, 4/5]$	$\emptyset \to \{abe\}[4, 4/5]$	$\emptyset \to \{abe\}[4, 4/5]$
	$\emptyset \to \{bcd\}[4, 4/5]$	$\emptyset \to \{bcde\}[4, 4/5]$	$\emptyset \to \{bcde\}[4, 4/5]$
		$\{abe\} \to \{cd\}[3, 3/4]$	$\{abe\} \to \{cd\}[3, 3/4]$

Definition 5.2.1. *Minimal non-redundant rules* $(\mathcal{MNR})$ are defined as:
$$\mathcal{MNR} = \{(r : X \to Y \backslash X) \in \mathcal{AR}|\neg\exists (r' : X' \to Y' \backslash X') \in \mathcal{AR}, r' \neq r \wedge$$
$$X' \subseteq X \wedge Y' \supseteq Y \wedge sup(r') = sup(r) \wedge conf(r') = conf(r)\}.$$
Generic basis for strong certain association rules $(\mathcal{GB})$ is defined as:
$$\mathcal{GB} = \{X \to Y \backslash X | Y \in FC \wedge X \in \mathcal{G}(Y) \wedge X \neq Y\}.$$
Informative basis for strong approximate association rules $(\mathcal{IB})$ is defined as:
$$\mathcal{IB} = \{r : X \to Y \backslash X | Y \in FC \wedge X \in \mathcal{G} \wedge \gamma(X) \subset Y \wedge conf(r) > minConf\}.$$

Transitive reduction of informative basis for approximate association rules, denoted by $\mathcal{RI}$, is defined as:

$\mathcal{RI} = \{(X \rightarrow Y \backslash X) \in \mathcal{IB} | \gamma(X)$ is a maximal proper subset of Y in $\mathcal{FC}\}$.   ◇

Clearly, $\mathcal{RI} \subseteq \mathcal{IB}$. The rules in $\mathcal{IB}$ that are not present in $\mathcal{RI}$ can be retrieved by applying confidence transitivity inference [10]. Hence, $\mathcal{RI}$ is equivalent to $\mathcal{IB}$. In further considerations we will concentrate on $\mathcal{IB}$. As stated in [3,10], all strong certain association rules can be derived solely from $\mathcal{GB}$.

Proposition 5.2.1 [3,10].
a) $\mathcal{GB} \subseteq \mathcal{MNR}$.
b) $(\mathcal{GB}, CCI)$ is a lossless and informative representation of strong certain rules.
c) $\mathcal{IB} \subseteq \mathcal{MNR}$.

It was also argued in [3,10] that $(\mathcal{IB}, CCI)$ is a lossless and informative representation of strong approximate rules. Unfortunately, this claim is not correct, that is $(\mathcal{IB}, CCI)$ is neither lossless nor informative representation of strong approximate rules. We will prove that the claim is invalid by showing a weak point in its proof delivered in [3,10]. It is stated there, that for a strong approximate rule $X \rightarrow Y \backslash X$, there is a frequent generator $X' \subseteq X$ such that $\gamma(X') = \gamma(X)$ and there is a frequent closed itemset $Y' \supseteq Y$ such that $Y' = \gamma(Y)$. Hence, $X \rightarrow Y \backslash X$ has a corresponding rule $X' \rightarrow Y' \backslash X'$ in $\mathcal{IB}$ such that $sup(X \rightarrow Y \backslash X) = sup(X' \rightarrow Y' \backslash X')$ and $conf(X \rightarrow Y \backslash X) = conf(X' \rightarrow Y' \backslash X')$. Clearly, in order to determine support and confidence of $X \rightarrow Y \backslash X$, we need to know how to identify $X' \rightarrow Y' \backslash X'$ among other rules in $\mathcal{IB}$. In the context of closure-closure rule inference, the problem consists in determining the closures $\gamma(X), \gamma(Y)$ as well as $\gamma(X")$ for each rule $X" \rightarrow Y" \backslash X"$ in $\mathcal{IB}$. We do not need to determine the closure of $Y"$ because $Y"$ is a closed itemset as a base of a rule in $\mathcal{IB}$, and by definition is equal to $\gamma(Y")$. Property 3.4.2 was offered in [3,10] as a method of determining closures of itemsets. Unfortunately, the property does not guarantee the correct determination of closures of frequent itemsets, when some frequent closed itemsets are not known. In the case of $\mathcal{IB}$, the only known frequent closed itemsets are the bases of the rules in $\mathcal{IB}$. Nevertheless, there is no guarantee that the set of closures of frequent generators that are antecedents of rules in $\mathcal{IB}$ is contained in the known frequent closed itemsets! We will prove this observation by means of the example below showing a case when the closure of an antecedent of a rule in $\mathcal{IB}$ is not derivable from $\mathcal{IB}$ itself:

Example 5.2.1. Let $\mathcal{D}$ be the database from Table 1. Given $minSup = 2$ and $minConf = 77\%$, one would obtain: $\mathcal{IB} = \{\emptyset \rightarrow \{abe\}[4, 4/5], \emptyset \rightarrow \{bcde\}[4, 4/5]\}$ (see Table 3). Both rules in $\mathcal{IB}$ have $\emptyset$ as an antecedent. We claim that determining $\gamma(\emptyset)$ from $\mathcal{IB}$ is not feasible. The only known closed itemsets implied by $\mathcal{IB}$ are $\{abe\}$ and $\{bcde\}$, which are the bases of the respective rules in $\mathcal{IB}$. The trial to determine $\gamma(\emptyset)$ as the smallest superset among known closed itemsets (following Property 3.4.2) fails, as both $\{abe\}$ and $\{bcde\}$ are minimal supersets of $\emptyset$. Actually, the real closure of $\emptyset$ is equal to $\{be\}$, but this information is not derivable from $\mathcal{IB}$.   □

The possible inability to determine closures of itemsets correctly may hence lead to the inability to derive some strong rules or to identifying weak rules as strong ones. The problem can be solved partially by applying the cover operator that was used so far for reasoning with the $\mathcal{RR}$ representation. Below we provide our newest results, which, in particular, will lead us to the conclusion that $(\mathcal{IB}, \mathcal{C})$ is a lossless representation of strong approximate rules. Let us start with comparing minimal non-redundant association rules with the $\mathcal{RR}$ representation.

Lemma 5.2.1. $\mathcal{MNR} = \{r \in \mathcal{AR} | \neg \exists r' \in \mathcal{AR}, r' \neq r \wedge r \in C(r')$
$\wedge \, sup(r') = sup(r) \wedge conf(r') = conf(r)\}$.
Proof. Follows immediately by definition of $\mathcal{MNR}$ and Proposition 3.3.3.   □

The lemma above states that a minimal non-redundant rule does not belong to the cover of another rule having the same support and confidence, although it may belong to the cover of a rule of different support or confidence. On the other hand, a representative rule does not belong to the cover of any rule irrespectively its support and confidence. Therefore, a representative rule is minimal non-redundant, but not necessarily vice versa.

Corollary 5.2.1. $\mathcal{RR} \subseteq \mathcal{MNR}$.

We claim that each strong association rule is covered by some minimal non-redundant association rule of the same support and confidence:

Theorem 5.2.1. $\forall r \in \mathcal{AR} \, \exists r' \in \mathcal{MNR} \, (r \in C(r') \wedge sup(r) = sup(r') \wedge conf(r) = conf(r'))$.
Proof. Let $r_0 \in \mathcal{AR}$. By Lemma 5.2.1, $r_0 \in \mathcal{MNR}$ or there is $r \in \mathcal{AR}$ such that $r \neq r_0, r_0 \in C(r)$, $sup(r) = sup(r_0)$ and $conf(r) = conf(r_0)$ (*). By Corollary 3.3.1, if $r \neq r_0$ and $r_0 \in C(r)$, then $r \notin C(r_0)$ (**). Now, we will consider two cases: 1) $r_0 \in \mathcal{MNR}$, 2) $r_0 \notin \mathcal{MNR}$.
Ad. 1) If $r_0 \in \mathcal{MNR}$, then $r_0 \in C(r_0)$ by definition of the cover operator, and the theorem holds.
Ad. 2) By (*), (**), and by the fact that the number of rules in $\mathcal{AR}$ is finite, there is a non-empty sequence of rules $< r_1, \ldots, r_n >$ in $\mathcal{AR}$ such that for all $i, j = 0..n$, $i < j$, the following holds: $r_i \neq r_j, r_i \in C(r_j), r_j \notin C(r_i), sup(r_i) = sup(r_j) = sup(r_0), conf(r_i) = conf(r_j) = conf(r_0)$, and r_n is not covered by another strong association rule of the same support and confidence. Hence, $r_n \in \mathcal{MNR}$ and $sup(r_n) = sup(r_0), conf(r_n) = conf(r_0)$. Now, by Proposition 3.3.4 (of cover transitivity), $r_0 \in C(r_n)$, which ends the proof.   □

Next, we observe that the union of $\mathcal{GB}$ and $\mathcal{IB}$ constitutes $\mathcal{MNR}$.
Lemma 5.2.2.
a) $\mathcal{GB} = \{X \to Y \backslash X | Y \in FC \wedge X \in FG \wedge X \subset Y \wedge \gamma(X) = Y\}$.
b) $\mathcal{IB} = \{r : X \to Y \backslash X | Y \in FC \wedge X \in FG \wedge X \subset Y \wedge \gamma(X) \neq Y \wedge$
   $conf(r) > minConf\}$,
c) $\mathcal{GB} \cup \mathcal{IB} = \{r : X \to Y \backslash X | Y \in FC \wedge X \in FG \wedge X \subset Y \wedge$
   $conf(r) > minConf\}$,
d) $\mathcal{GB} \cup \mathcal{IB} = \mathcal{MNR}$,

e) $\mathcal{GB} = \{r \in \mathcal{MNR}| conf\,(r) = 1\}$,

f) $\mathcal{IB} = \{r \in \mathcal{MNR}| conf\,(r) < 1\}$.

Proof. Ad. a) Follows the definition of a generator.

Ad. b) Let $Y \in FC$. By Property 2.1d, $\gamma\,(X) \subseteq Y$ iff $X \subseteq Y$. Thus, $\gamma\,(X) \subset Y$ iff $\gamma\,(X) \neq Y$ and $X \subseteq Y$ iff $\gamma\,(X) \neq Y$ and $X \subset Y$ (by that fact that $X \subseteq \gamma\,(X)$). Hence, $\mathcal{IB} = \{X \rightarrow Y\backslash X|Y \in FC \wedge X \in FG \wedge X \subset Y \wedge \gamma\,(X) \neq Y \wedge conf\,(r) > minConf\}$ (each X is frequent as a subset of Y, which is frequent).

Ad. c) Follows by Lemma 5.2.2a-b and Property 2.1a implying that $Y \in FC \wedge X \subset Y \wedge \gamma\,(X) = Y$ is equivalent to $Y \in FC \wedge X \subset Y \wedge sup\,(X) = sup\,(Y)$.

Ad. d) By Proposition 5.2.1a,c, $\mathcal{GB} \cup IB \subseteq \mathcal{MNR}$. Now, it is sufficient to prove that each minimal non-redundant association rule belongs to $\mathcal{GB} \cup \mathcal{IB}$. The proof will be made by contradiction: Let $(r : X \rightarrow Y\backslash X) \in \mathcal{MNR}\backslash(\mathcal{GB} \cup \mathcal{IB})$. Hence, by Lemma 5.2.2c, $X \notin FG$ or $Y \notin FC$. Let X' be a subset of X such that $X' \in \mathcal{G}\,(\gamma\,(X))$, and $Y' \in \gamma\,(Y)$. Then, $r' : X' \rightarrow Y'\backslash X'$ is a strong association rule such that $r' \neq r, sup\,(r') = sup\,(r), conf\,(r') = conf\,(r), Y \subseteq Y'$, and $X \supseteq X'$. Thus, $r \in \mathcal{MNR}$, which contradicts the assumption.

Ad. e) Follows by Lemma 5.2.2a,c,d and Property 2.1a implying that $Y \in FC \wedge X \subset Y \wedge \gamma\,(X) = Y$ is equivalent to $Y \in FC \wedge X \subset Y \wedge sup\,(X) = sup\,(Y)$.

Ad. f) Follows by Lemma 5.2.2b-d and the fact that $Y \in FC \wedge X \subset Y \wedge \gamma\,(X) \neq Y$ is equivalent to $Y \in FC \wedge X \subset Y \wedge sup\,(X) > sup\,(Y)$. $\qquad\square$

Lemma 5.2.2 allows us to infer that $\mathcal{GB} \cup \mathcal{IB}$ constitutes a superset of $\mathcal{RR}$ and $\mathcal{GB}$ itself is a superset of certain rules in $\mathcal{RR}$s:

Corollary 5.2.2.

a) $\mathcal{GB} \cup \mathcal{IB} \supseteq \mathcal{RR}$,

b) $\mathcal{GB} \supseteq \{r \in \mathcal{RR}| conf\,(r) = 1\}$.

Proof. Ad. a) Follows immediately by Lemma 5.2.2d and Corollary 5.2.1.

Ad. b) By Corollary 5.2.2a, $\mathcal{GB} \cup \mathcal{IB} \supseteq \{r \in \mathcal{RR}| conf\,(r) = 1\}$. On the other hand, by Corollary 5.2.1 and Lemma 5.2.2f, $\mathcal{IB} \cap \{r \in \mathcal{RR}| conf\,(r) = 1\} = \emptyset$. Hence, $\mathcal{GB} \supseteq \{r \in \mathcal{RR}| conf\,(r) = 1\}$. $\qquad\square$

Table 3 illustrates this corollary.

Proposition 5.2.2.

a) $\bigcup_{r \in \mathcal{GB} \cup \mathcal{IB}} C\,(r) = \mathcal{AR}$,

b) $\bigcup_{r \in \mathcal{GB}} C\,(r) = \{X \rightarrow Y\backslash X \in \mathcal{AR}| conf\,(X \rightarrow Y\backslash X) = 1\}$,

c) $\bigcup_{r \in \mathcal{IB}} C\,(r) \supseteq \{X \rightarrow Y\backslash X \in \mathcal{AR}| 1 > conf\,(X \rightarrow Y\backslash X) > minConf\}$.

Proof. Ad. a) The fact that $\mathcal{GB} \cup \mathcal{IB}$ is a superset of $\mathcal{RR}$, implies $\bigcup_{r \in \mathcal{GB} \cup \mathcal{IB}} C\,(r) \supseteq \bigcup_{r \in \mathcal{RR}} C\,(r) = \mathcal{AR}$. On the other hand, $\mathcal{GB} \cup \mathcal{IB}$ contains only strong rules, so their covers also contain only strong rules (by Proposition 3.3.1a). Hence, $\bigcup_{r \in \mathcal{GB} \cup \mathcal{IB}} C\,(r) = \mathcal{AR}$.

Ad. b) Let r_0 be a rule in $\mathcal{AR}$ such that $conf\,(r_0) = 1$. Then, by Theorem 5.2.1, there is r in $\mathcal{MNR}$ such that $r_0 \in C\,(r)$ and $conf\,(r) = 1$. By Lemma 5.2.2e, we conclude further, r_0 is covered by some rule r in $\mathcal{GB}$. More generally, each certain rule is covered by some rule in $\mathcal{GB}$; that is $\bigcup_{r \in \mathcal{GB}}(C\,(r) \supseteq$

$(\{X \to Y \backslash X\} \in \mathcal{AR} | conf(X \to Y \backslash X) = 1\})(*)$. On the other hand, for all $r \in \mathcal{GB}$ $(C(r) \subseteq \{(X \to Y \backslash X) \in \mathcal{AR} | conf(X \to Y \backslash X) = 1\})$ by Lemma 5.2.2e and Proposition 3.3.1a $(**)$. The proposition follows immediately by $(*)$ and $(**)$.

Ad. c) Follows by Proposition 5.2.2a-b. $\qquad\qquad\qquad\qquad\qquad\qquad\qquad\square$

Hence, $(\mathcal{GB} \cup \mathcal{IB}, C)$ (or equivalently, $(\mathcal{MNR}, C)$) is a lossless and sound representation of $\mathcal{AR}$ (by Proposition 5.2.2a), and $(\mathcal{GB}, C)$ is a lossless and sound representation of strong certain association rules (by Proposition 5.2.2b). By Proposition 5.2.2c, $(\mathcal{IB}, C)$ is a lossless representation of strong approximate association rules (as it derives all strong approximate rules), although potentially not sound (as it may derive also strong certain association rules). In fact, $(\mathcal{IB}, C)$ is not sound in general, which proves the example below.

Example 5.2.2. Let $\mathcal{D}$ be the database from Table 1. Given $minSup = 2$ and $minConf = 77\%$, one would obtain: $\mathcal{IB} = \{\emptyset \to \{abe\}[4, 4/5], \emptyset \to \{bcde\}[4, 4/5]\}$ (see Table 3). Let us trace the evaluation of the rule $\{bc\} \to \{d\}$ with this $\mathcal{IB}$. Here, $\{bc\} \to \{d\}$ belongs to the cover of only one rule in $\mathcal{IB}$, namely: $\emptyset \to \{bcde\}$. Thus, the rule $\{bc\} \to \{d\}$ is identified as strong approximate with support equal to $sup(\emptyset \to \{bcde\}) = 4$ and confidence equal to $conf(\emptyset \to \{bcde\}) = 4/5$. This identification is not correct since $\{bc\} \to \{d\}$ is a certain rule, which is covered by certain rule $\{c\} \to \{bde\}[4, 1]$ in $\mathcal{GB}$ (see Table 3). Clearly, if we used all rules in both $\mathcal{GB}$ and $\mathcal{IB}$ for evaluating $\{bc\} \to \{d\}$, we would not make such a mistake. Undoubtedly, $\{bc\} \to \{d\}$ belongs to the cover of both approximate rule: $\emptyset \to \{bcde\}$ and certain rule: $\{c\} \to \{bde\}$, nevertheless the confidence of $\{bc\} \to \{d\}$ equals to the maximum from the confidences of the two covering rules; i.e. $conf(\{bc\} \to \{d\}) = conf(\{c\} \to \{bde\}) = 1$. $\quad\square$

In the next proposition, we specify how to calculate support and confidence of any strong association rule based on the rules in $\mathcal{GB}$ and/or $\mathcal{IB}$ that cover the evaluated rule.

Proposition 5.2.3.

a) $\forall r \in \mathcal{AR}, sup(r) = max\{sup(r) | r \in C(r'), r' \in (\mathcal{GB} \cup \mathcal{IB})\}$ and
   $conf(r) = max\{sup(r) | r \in C(r'), r' \in (\mathcal{GB} \cup \mathcal{IB})\}$,

b) $\forall r \in \mathcal{AR}, conf(r) = 1$ implies $sup(r) = max\{sup(r) | r \in C(r'), r' \in \mathcal{GB}\}$,

c) $\forall r \in \mathcal{AR}, conf(r) < 1$ implies $sup(r) = max\{sup(r) | r \in C(r'), r' \in \mathcal{IB}\}$
   and $conf(r) = max\{sup(r) | r \in C(r'), r' \in \mathcal{IB}\}$.

Proof. Ad. a) Let $r \in \mathcal{AR}$. By Proposition 3.3.1a, if there are several rules in $\mathcal{GB} \cup \mathcal{IB}$ covering r, then support and confidence of r is not less than supports and confidences of all those covering rules. By Theorem 5.2.1 and Lemma 5.2.2d, for each rule r in $\mathcal{AR}$, there is a covering rule in $\mathcal{GB} \cup \mathcal{IB}$ of the same support and confidence as those of r. Hence, support and confidence of r is equal to maximum from supports and confidences, respectively, of all rules in $\mathcal{GB} \cup \mathcal{IB}$ that cover r.

Ad. b) By Proposition 5.2.3a, $\forall r \in \mathcal{AR}, conf(r) = 1$ implies $sup(r) = max\{sup(r) | r \in C(r'), r' \in (\mathcal{GB} \cup \mathcal{IB})\}$. We are now to prove that $\mathcal{GB}$ itself is sufficient to determine supports of certain rules: Let $(r : X \to Y \backslash X) \in \mathcal{AR}$ and $conf(r) = 1$. Let $r' : X' \to Y' \backslash X'$ be a rule such that $X' \subseteq X, X' \in \mathcal{G}(\gamma(X))$

and $Y' = \gamma(Y)$. Hence, $X' \subseteq X \subset Y \subseteq Y'$, and $sup(X') = sup(X) = sup(Y) = sup(Y')$ (*). Inferring further, $X' \neq Y'$ and $r \in C(r')$ (**). Since r is strong, then Y is frequent, and hence $Y' \in FC$ (***). By (*) and Property 2.1a, $\gamma(X') = \gamma(Y')$ and by (***) $\gamma(Y') = Y'$. Since, X' is a generator and $\gamma(X') = Y'$, then $X' \in G(Y')$ (****). By (*), (**), (****) and (***), $r' \in GB, r \in C(r')$ and $sup(r) = sup(r')$. Hence, we proved that for any certain strong association rule r there is always a covering rule r' in GB of the same support.

Ad. c) By Proposition 5.2.3a, each strong association rule r has covering rules in $GB \cup IB$ that determine its support and confidence. Nevertheless, an approximate rule (whose confidence is less than 1) cannot be covered by any certain rule (whose confidence is equal to 1) according to Property 3.3.1a. For this reason, no rule in GB, which contains only certain rules, covers any approximate rule. By this observation and Proposition 5.2.3a, $sup(r) = max\{sup(r) | r \in C(r'), r' \in IB\}$ and $conf(r) = max\{sup(r) | r \in C(r'), r' \in IB\}$.   □

Propositions 5.2.2, 5.2.3 lead us to the following conclusion:

Corollary 5.2.3.
a) $(GB \cup IB, C)$ is a lossless, sound and informative representation of AR.
b) (GB, C) is a lossless, sound and informative representation of strong certain association rules.
c) $(GB \cup IB, C)$ is a lossless, sound and informative representation of strong approximate association rules.

Proof. Ad. a) By Propositions 5.2.2a and 5.2.3a.
Ad. b) By Propositions 5.2.2b and 5.2.3b.
Ad. c) Follows immediately by Corollary 5.2.3a-b: $(GB \cup IB, C)$ derives each strong association rule as well as its support and confidence. If the confidence of the rule is equal to 1, then the rule is certain. Otherwise it is approximate.   □

Now, let us return for a while to the RI representation. Since, IB can be obtained from RI by confidence transition property (CTP), we conclude:
- (RI, C) is a lossless representation of strong approximate association rules, although in general it is neither sound nor informative.
- $(RI \cup IB, C)$ is a lossless, sound and informative representation of strong approximate association rules.

One cannot state on theoretical basis what is more concise, RI or RR. Table 3, illustrates cases, when RR is less concise than RI $(minConf = 70\%)$, the same as RI $(minConf = 77\%)$, and more concise than RI $(minConf = 50\%)$, depending on the confidence threshold value. On the other hand, $GB \cup RI$, which derives both strong certain and strong rules correctly, is more numerous than RR in each case in our example.

5.3   Duquenne-Guigues Basis, Proper Basis, and Related Representations

In this subsection, we recall after [10,11] a concept of pseudo-closed itemsets and related representations of association rules based on pseudo- and closed itemsets.

Next we examine these concepts more thoroughly showing their strong and weak points and proposing solutions to the weak points:

Definition 5.3.1. An *itemset* $X \subseteq I$ *is pseudo-closed* iff $\gamma(X) \neq X$ and $\forall Y \subset X$, such that Y is a pseudo-closed itemset, we have $\gamma(Y) \subset X$. The set FP of frequent pseudo-closed itemsets is defined as:

$$FP = \{X \in \mathcal{F} | \gamma(X) \neq X \text{ and } \forall Y \in FP \text{ such that } Y \subset X, \gamma(Y) \subset X\}.$$

Duquenne-Guigues basis for certain association rules $(\mathcal{DG})$ is defined as:

$$DG = \{r : X \to \gamma(X) \setminus X | X \in FP\}.$$

Proper basis for approximate association rules $(\mathcal{PB})$ is defined as:

$$\mathcal{PB} = \{r : X \to Y \setminus X | X, Y \in FC \wedge X \subset Y \wedge conf(r) > minConf\}.$$

Luxenburger transitive reduction of proper basis for approximate association rules, denoted by $\mathcal{LB}$, is defined as [10, 11]:

$$\mathcal{LB} = \{(X \to Y \setminus X) \in \mathcal{PB} | X \text{ is a maximal proper subset of } Y \text{ in } FC\}.$$

Structural basis for strong approximate association rules $(\mathcal{SB})$ is a subset of the informative basis and is defined as follows:

$$\mathcal{SB} = \{(X \to Y \setminus X) \in \mathcal{PB} | \neg \exists (X' \to Y \setminus X') \in \mathcal{PB}, X' \neq X \wedge$$
$$X' \text{precedes } X \text{ in the lexicographic order}\}. \qquad \diamond$$

Clearly, $\mathcal{SB} \subseteq \mathcal{LB} \subseteq \mathcal{PB}$. As stated in [10], all strong certain association rules can be derived solely from $\mathcal{DG}$ and all strong approximate association rules can be derived solely from $\mathcal{PB}$. The certain association rules are derived from $\mathcal{GB}$ by applying Armstrong's axioms. In the case of $\mathcal{SB}$, the bases of rules in $\mathcal{SB}$ are distinct closed itemsets. The rules in $\mathcal{PB}$ that are not present in $\mathcal{LB}$ or $\mathcal{SB}$ can be retrieved by applying confidence transitivity property (Property 3.2.1) [10].

Proposition 5.3.1 [10].

a) $(\mathcal{DG}, AA)$ is a lossless and non-informative representation of strong certain rules (confidences of derivable rules are equal to 1, but supports are not determinable).

b) There is no smaller set of rules than $\mathcal{DG}$ from which one could derive all strong certain association rules.

c) $(\mathcal{PB}, CCI)$ is a lossless representation of strong approximate rules.

It is also argued in [10] that $(\mathcal{PB}, CCI)$ is an informative representation of strong approximate rules. We disagree with this claim. [3] However, before we will prove this, let us provide an example illustrating reasoning with $(\mathcal{DG}, AA)$:

Example 5.3.1. Given $minSup = 2$, one finds the following pseudo-closed itemsets for database from Table 1: $FP = \{\emptyset, \{bce\}, \{bde\}\}$. Their closures are $\{be\}$, $\{bcde\}$, and $\{bcde\}$, respectively. Hence: $\mathcal{DG} = \{\emptyset \to \{be\}, \{bce\}rightarrow\{d\},$ $\{bde\} \to \{c\}\}$ (see Table 3). We will demonstrate the use of $\mathcal{DG}$ for evaluating rule $\{a\} \to \{be\}$ by means of Armstrong's axioms. Clearly, Armstrong's axioms applied to rule $\emptyset \to \{be\}$ in $\mathcal{DG}$ derive $\{a\} \to \{be\}$ as a certain rule. Nevertheless,

---

[3] It is stated in [11], that $(\mathcal{LB}, CCI)$ is an informative representation of strong approximate rules. Nonetheless, we disagree with that claim either.

there is no possibility to determine support of $\{a\} \to \{be\}$, since $\mathcal{DG}$ does not contain any rule with antecedent or base equal to $\{abe\}$.   □

When support of a derived certain rule is unknown, one is not able to determine if the rule is strong or not. Hence, we conclude $(\mathcal{DG}, \mathcal{AA})$ is a lossless, although not sound and non-informative, representation of strong certain association rules. [4] It however derives only certain rules.

Presented in the literature [10,11] mechanism of reasoning with $\mathcal{PB}$ encounters the same problems as we faced when discussing reasoning with the $\mathcal{GB}$ and $\mathcal{IB}$ representations: It is stated in [10,11], that for a strong approximate rule $X \to Y\backslash X$, there are frequent closed itemsets $X' \subseteq X$ and $Y' \supseteq Y$, such that $X' = \gamma(X)$ and $Y' = \gamma(Y)$. Hence, $X \to Y\backslash X$ has a corresponding rule $X' \to Y'\backslash X'$ in $\mathcal{PB}$ such that $sup(X \to Y\backslash X) = sup(X' \to Y'\backslash X')$ and $conf(X \to Y\backslash X) = conf(X' \to Y'\backslash X')$. The rule $X' \to Y'\backslash X'$ can be identified among other rules in $\mathcal{PB}$ by determining X' and Y' as the respective closures according to Property 3.4.2. Nevertheless, it has not been considered in [10,11] how to evaluate rules that are not strong. Below we prove by means of an example that $(\mathcal{PB}, CCI)$ is neither sound nor informative representation of strong approximate rules.

Example 5.3.2. Let $\mathcal{D}$ be the database from Table 4:

Table 4. Another example database $\mathcal{D}$

Id	Transaction
T_1	$\{abc\}$
T_2	$\{abc\}$
T_3	$\{abc\}$
T_4	$\{ab\}$
T_5	$\{a\}$
T_6	$\{a\}$

For $minSup = 2$, we have the following frequent generators and closed itemsets:
$$FG = \{\emptyset_{[6]}, \{b\}_{[4]}, \{c\}_{[3]}\},$$
$$FC = \{\{a\}_{[6]}, \{ab\}_{[4]}, \{abc\}_{[3]}\}.$$
The following representations will be extracted for $minConf = 70\%$:
$$\mathcal{DG} = \{\emptyset \to \{a\}[6, 6/6], \{ac\} \to \{b\}[3, 3/3]\}, \text{ and}$$
$$\mathcal{PB} = \{\{ab\} \to \{c\}[3, 3/4]\}.$$

Let us now evaluate a candidate rule $\{a\} \to \{c\}$ using only $\mathcal{PB}$: Since $\{ab\} \to \{c\}$ is the only rule in $\mathcal{PB}$, $\{abc\}$ is the smallest closed itemset known in $\mathcal{PB}$, which is a superset of the candidate rule base $\{ac\}$, and $\{ab\}$ is the smallest closed itemset known in $\mathcal{PB}$, which is a superset of the rule antecedent $\{a\}$ (here, $\mathcal{PB}$ falsely implies that $\gamma(\{a\}) = \{ab\}$, while in reality $\gamma(\{a\}) = \{a\}$!). Hence, $\{a\} \to \{c\}$ is identified as a strong approximate rule with support equal to $sup(\{ab\} \to \{c\}) = 3$ and $conf(\{ab\} \to \{c\}) = 3/4$. Nevertheless, looking at Table 4, one can easily check that $conf(\{a\} \to \{c\}) = 3/6 < minConf$. Thus, $\{a\} \to \{c\}$ was incorrectly identified with $(\mathcal{PB}, CCI)$ as strong approximate.   □

The example above proves that $(\mathcal{PB}, CCI)$ (and by this $(\mathcal{LB}, \{CCI, CTP\})$ and $(\mathcal{LB}, \{SB, CTP\})$) is neither sound nor informative representation of strong approximate rules and may derive rules with insufficient confidence. We claim however it is possible to turn $(\mathcal{PB}, CCI)$ (and by this $(\mathcal{LB}, \{CCI, CTP\})$ and

---

[4] It is proposed in [11] to augment $(\mathcal{DG}, \mathcal{AA})$ with the set of all infrequent pseudo-clased itemsets (IP) to suppres derivation of certain rules that are not strong.

$(\mathcal{LB}, \{SB, CTP\}))$ into sound and informative representation by extending it with $\mathcal{DG}$.

We observe first that, although in general $\mathcal{DG}$ is not sufficient to determine support and closures of itemsets, it is however sufficient to determine whether a given frequent closed itemset Y is a closure of an itemset X.

Proposition 5.3.2. Let X, Y be itemsets such that $X \subset Y$ and $Y \in \mathcal{F}$.
a) $X \to Y \backslash X$ is derivable from $(\mathcal{DG}, AA)$ iff $sup(X) = sup(Y)$,
b) $X \to Y \backslash X$ is derivable from $(\mathcal{DG}, AA)$ iff $\gamma(X) = \gamma(Y)$,
c) $Y \in \mathcal{C} \wedge X \to Y \backslash X$ is derivable from $(\mathcal{DG}, AA)$ iff $\gamma(X) = Y$.
Proof. Ad. a) By Proposition 5.3.1a.
Ad. b) By Proposition 5.3.2a and Property 2.1a.
Ad. c) $Y \in \mathcal{C}$ iff $\gamma(Y) = Y$. Inferring further by Proposition 5.3.2b, $Y \in \mathcal{C}$ and $X \to Y \backslash X$ is derivable from $(\mathcal{DG}, AA)$ iff $\gamma(Y) = Y \wedge \gamma(X) = \gamma(Y)$. Hence, $Y \in \mathcal{C}$ and $X \to Y \backslash X$ is derivable from $(\mathcal{DG}, AA)$ iff $\gamma(X) = Y$.   □

We will use the proposition above to correct the rule validation procedure from Example 5.3.2 by additional use of $\mathcal{DG}$:

Example 5.3.3. Let $\mathcal{D}$ be the database from Table 4. $\{a\} \to \{c\}$ is an evaluated rule. As shown in Example 5.3.2, $\{ab\} \to \{c\}$ is the only rule in $\mathcal{PB}$ that might be equal to $\gamma(\{a\}) \to \gamma(\{ac\}) \backslash \gamma(\{a\})$. We can use $\mathcal{DG}$ in order to test if $\gamma(\{a\}) = \{ab\}$ and $\gamma(\{ac\}) = \{abc\}$. One may note that $(\mathcal{DG}, AA)$ does not derive $\{a\} \to \{ab\}$. (Among rules with $\{a\}$ in the antedecedent that are derivable from $\mathcal{DG}$ by Armstrong's axioms, only $\{a\} \to \{a\}$ is a rule with maximal base.) Hence, by Proposition 5.3.2c, $\gamma(\{a\}) \neq \{ab\}$. Thus, $\gamma(\{a\}) \to \gamma(\{ac\}) \backslash \gamma(\{a\})$ is not equal to the $\mathcal{PB}$'s rule: $\{ab\} \to \{c\}$. This way we made certain $\mathcal{PB}$ does not contain $\gamma(\{a\}) \to \gamma(\{ac\}) \backslash \gamma(\{a\})$, and by this $\{a\} \to \{c\}$ should not be identified as strong.   □

Example 5.3.3 indicates the following way of determining for any rule if it is strong approximate or not by means of the closure-closure rule inference and Armstrong's axioms applied to $\mathcal{PB}$ and $\mathcal{DG}$: Let $X \to Y \backslash X$ be a rule under evaluation. Compare the rule $X \to Y \backslash X$ with the rules in $\mathcal{PB}$ as long as either 1) some rule $X' \to Y' \backslash X'$ is found in $\mathcal{PB}$ such that $\gamma(X) = X'$ and $\gamma(Y) = Y'$, or 2) there is no such rule in $\mathcal{PB}$. In the former case, checking whether $\gamma(X) = X'$ and $\gamma(Y) = Y'$ can be carried out by checking if $X \to X' \backslash X$ and $Y \to Y' \backslash Y$ are derivable by Armstrong's axioms from $\mathcal{DG}$ (by Proposition 5.3.2c). If so, then $X \to Y \backslash X$ is a strong approximate association rule (by the closure-closure rule inference). In the latter case, $X \to Y \backslash X$ is not strong approximate.

Corollary 5.3.1. $(\mathcal{PB} \cup \mathcal{DG}, \{CCI, AA\})$ (and by this $(\mathcal{LB} \cup \mathcal{DG}, \{CCI, AA\})$ and $(SB \cup \mathcal{DG}, \{CCI, AA\})$) is a lossless, sound and informative representation of $\mathcal{AR}$.

There is no strict relationship between $\mathcal{DG}, \mathcal{PB}, \mathcal{LB}, SB$, and the $\mathcal{RR}$ representation. As shows Table 3, the number of rules in $\mathcal{DG}, \mathcal{PB}, \mathcal{LB}$, or SB, respectively, can be smaller or greater than the number of rules in $\mathcal{RR}$.

5.4   Summary of Association Rules Representations

In this section we summarize the properties of the reviewed representations.

Table 5. The $\mathcal{RR}$ and closed/pseudo-closed itemsets representations for database from Table 1 given $minSup = 2$ and $minConf = 77\%$, 70% and 50%, respectively

Representation	Intended derivable rules	Inference mechanism	Lossless	Sound	Informative
$\mathcal{RR}$	$\mathcal{AR}$	C	yes	yes	no
$\mathcal{GB}$	certain $\mathcal{AR}$	C	yes	yes	yes
$\mathcal{IB}$	approximate AR	C	yes	no*	no
$\mathcal{RI}$	approximate AR	C, CTP	yes	no*	no
$\mathcal{IB}$ using $\mathcal{GB}$	approximate $\mathcal{AR}$	C	yes	yes	yes
$\mathcal{RI}$ using $\mathcal{GB}$	approximate $calAR$	C, CTP	yes	yes	yes
$\mathcal{DG}$	certain $\mathcal{AR}$	AA	yes	no**	no
$\mathcal{PB}$	approximate $\mathcal{AR}$	CCI	yes	no	no***
$\mathcal{LB}$ (or $\mathcal{SB}$)	approximate $\mathcal{AR}$	CCI, CTP	yes	no	no***
$\mathcal{PB}$ using $\mathcal{DG}$	approximate $\mathcal{AR}$	AA, CCI	yes	yes	yes
$\mathcal{LB}$ (or $\mathcal{SB}$) using $\mathcal{DG}$	approximate $\mathcal{AR}$	AA, CCI, CTP	yes	yes	yes

(*) Cover operator derives not only strong approximate association rules from $\mathcal{IB}$ ($\mathcal{RI}$), but also strong certain ones.

(**) Armstrong's axioms derive only certain rules from $\mathcal{DG}$, but in general it is not known which of them are strong or not, since their supports may be not derivable.

(***) Support and confidence is determined properly for strong association rules. For other rules that are derivable from $\mathcal{PB}$ ($\mathcal{LB}, \mathcal{SB}$), support or confidence is calculated improperly.

All the presented representations contain only rules whose bases are frequent closed itemsets and whose antecedents are frequent generators ($\mathcal{RR}, \mathcal{MNR},$ $\mathcal{GB}, \mathcal{IB}, \mathcal{RI}$) or closed itemsets ($\mathcal{PB}, \mathcal{LB}, \mathcal{SB}$) or pseudo-closed itemsets ($\mathcal{DG}$). Hence, all the representations except for $\mathcal{DG}$ are directly derivable from the lossless concise representations of frequent itemsets.

6   Conclusions

The paper was devoted to theoretical aspects of concise representations of association rules. We have pointed out that representations of strong association rules that are not sound are of no value even if lossless. We note that the notion of a representation of association rules must involve two parts: a set of rules and an associated inference mechanism. It is important to associate a suitable inference mechanism with each rule representation. In particular, we have shown that ($\mathcal{IB}, CCI$) (and by this ($\mathcal{RI}, \{CCI, CTP\}$)) is neither lossless, nor sound, nor informative representation of strong approximate rules, however $(\mathcal{IB}, C)$ (and by this $(\mathcal{RI}, \{C, CTP\})$) is lossless. In addition, we have shown that $(\mathcal{PB}, CCI)$ (and by this $(\mathcal{LB}, \{CCI, CTP\})$ and $(\mathcal{SB}, \{CCI, CTP\})$) is neither sound nor informative representation of strong approximate rules. We suggested to extend the representations of strong approximate rules (such as $(\mathcal{IB}, C)$ and $(\mathcal{PB}, CCI)$), which are not sound, with respective representations of strong certain rules $((\mathcal{GB}, C)$ and $(\mathcal{DG}, AA)$, respectively) in order to make the former ones sound and informative. Clearly, the disadvantage of this solution is increase in the size of the representations.

We have identified relationships between the representative rules and generic and informative basis. In particular, we have proved that the union of $\mathcal{GB}$ and $\mathcal{IB}$ is equal to $\mathcal{MNR}$ and always constitutes a superset of $\mathcal{RR}$. For a toy dataset, we have shown cases when $\mathcal{RR}$ is more concise than all the other representations (i.e. $\mathcal{MNR}, \mathcal{GB}, \mathcal{IB}, \mathcal{RI}, \mathcal{DG}, \mathcal{PB}, \mathcal{SB}, \mathcal{LB}$). Some counter-examples for the toy dataset were also provided.

References

1. Aggarwal C., Yu P.: On Line Generation of Association Rules. In: Proc. of ICDE-98, (1998) 402–411
2. Agrawal, R., Imielinski, T., Swami, A.: Mining Associations Rules between Sets of Items in Large Databases. In: Proc. of the ACM SIGMOD Conference on Management of Data. Washington, D.C. (1993) 207–216
3. Bastide Y., Pasquier N., Taouil R., Stumme G., Lakhal L.: Mining Minimal Non-redundant Association Rules Using Frequent Closed Itemsets. CL (2000) 972–986
4. Kryszkiewicz, M.: Representative Association Rules. In: Proc. of PAKDD '98. Melbourne, Australia. LNAI **1394**. Springer-Verlag (1998) 198–209
5. Kryszkiewicz M.: Closed Set Based Discovery of Representative Association Rules. In: Proc. of IDA'2001. September 13–15, Lisbon, Portugal. LNCS. Springer (2001) 350–359
6. Kryszkiewicz, M.: Concise Representation of Frequent Patterns based on Disjunction-free Generators. In: Proc. of ICDM '2001. San Jose, USA. IEEE Comp. SC. Press (2001) 305–312
7. Kryszkiewicz M., Concise Representations of Frequent Patterns and Association Rules. Habilitation dissertation, submitted to Warsaw University of Technology
8. Luong V.P., Representative Bases of Association Rules. In: Proc. of ICDM '2001. San Jose, USA. IEEE Comp. SC. Press (2001) 639–640
9. Luxenburger, M.: Implications Partielles dans un Contexte, Mathématiques, informatique et sciences humaines, 113, (1991) 35–55
10. Pasquier, N.: DM: Algorithmes d'extraction et de réduction des regles d'association dans les bases de données. Ph.D. Thesis. Univ. Pascal-Clermont-Ferrand II (2000)
11. Stumme G., Taouil R., Bastide Y., Pasquier N., Lakhal L.: Intelligent Structuring and Reducing of Association Rules with Formal Concept Analysis. KI/ÖGAI (2001), 335–350
12. Zaki, M.J.: Generating Non-redundant Association Rules. In: 6th ACM SIGKDD (2000)

Constraint-Based Discovery and Inductive Queries: Application to Association Rule Mining

Baptiste Jeudy and Jean-François Boulicaut

Institut National des Sciences Appliquées de Lyon
Laboratoire d'Ingénierie des Systèmes d'Information
Bâtiment Blaise Pascal
F-69621 Villeurbanne cedex, France
{Baptiste.Jeudy,Jean-Francois.Boulicaut}@lisi.insa-lyon.fr

Abstract. Recently inductive databases (IDBs) have been proposed to afford the problem of knowledge discovery from huge databases. Querying these databases needs for primitives to: (1) select, manipulate and query data, (2) select, manipulate and query "interesting" patterns (i.e., those patterns that satisfy certain constraints), and (3) cross over patterns and data (e.g., selecting the data in which some patterns hold). Designing such query languages is a long-term goal and only preliminary approaches have been studied, mainly for the association rule mining task. Starting from a discussion on the MINE RULE operator, we identify several open issues for the design of inductive databases dedicated to these descriptive rules. These issues concern not only the offered primitives but also the availability of efficient evaluation schemes. We emphasize the need for primitives that work on more or less condensed representations for the frequent itemsets, e.g., the (frequent) δ-free and closed itemsets. It is useful not only for optimizing single association rule mining queries but also for sophisticated post-processing and interactive rule mining.

1   Introduction

In the cInQ[1] project, we want to develop a new generation of databases, called *"inductive databases"* (IDBs), suggested by Imielinski and Mannila in [15] and formalized in, e.g., [10]. This kind of databases integrate *raw data* with *knowledge* extracted from *raw data*, materialized under the form of patterns into a common framework that supports the knowledge discovery process within a database framework. In this way, the process of KDD consists essentially in a querying process, enabled by an ad-hoc, powerful and universal query language that can deal either with raw data or patterns and that can be used throughout the whole KDD process across many different applications. We are far from an understanding of fundamental primitives for such query languages when considering various kinds of knowledge discovery processes. The so-called *association rule mining*

---

[1] This research is part of the cInQ project (IST 2000-26469) that is partially funded by the European Commission IST Programme – Future and Emergent Technologies.

D.J. Hand et al. (Eds.): Pattern Detection and Discovery, LNAI 2447, pp. 110–124, 2002.

process introduced in [1] has received a lot of attention these last five years and it provides an interesting context for studying the inductive database framework and the identification of promising concepts. Indeed, when considering this kind of local pattern, a few query languages can be considered as candidates like the MINE RULE operator [23], MSQL [16], and DMQL [14] (see also [5] for a critical evaluation of these three languages).

A *query language* for IDBs, is an extension of a query language that includes primitives for supporting every step of the mining process. When considering association rule mining, it means that the language enables to specify:

- The selection of data to be mined. It must offer the possibility to select (e.g., via standard queries but also by means of sampling), to manipulate and to query data and views in the database. Also, primitives that support typical preprocessing like quantitative value discretization are needed.
- The specification of the type of rules to be mined. It often concerns syntactic constraints on the desired rules (e.g., the size of the body) but also the specification of the sorts of the involved attributes.
- The specification of the needed background knowledge (e.g., the definition of a concept hierarchy).
- The definition of constraints that the extracted patterns must satisfy. Among others, this implies that the language allows the user to define constraints that specify the interestingness (e.g., using measures like frequency, confidence, etc) on the patterns to be mined.
- The post-processing of the generated results. The language must allow to browse the collection of patterns, apply selection templates, *cross over* patterns and data, e.g., by selecting the data in which some patterns hold, or to manipulate results with some aggregate functions.

The satisfaction of a *closure property*, i.e., the user queries an inductive database instance and the result is again an inductive database instance is crucial for supporting the dynamic aspect of a discovery process and its inherent interactivity. This closure property can be achieved by means of the storage of the extracted patterns in the same database.

Contribution. Relating the inductive database framework with constraint-based mining enables to widen the scope of interest of this framework to various contributions in the data mining and machine learning communities. Then, we consider that it is useful to emphasize the interest of several condensed representations for frequent patterns that have been studied the last three years. However, algorithms for mining these representations have been already published and will not be discussed here. Only a few important conceptual issues like regeneration or the need for constrained condensed representations are discussed. Last, we sketch why these concepts are useful not only for the optimization of a single association rule mining query but also for sophisticated rule post-processing and interactive association rule mining.

Organization of the paper. First, we discuss the MINE RULE proposal to identify several open issues for the design of inductive databases dedicated to association rule mining (Section 2). It concerns not only the offered primitives but also the availability of efficient evaluation schemes. In Section 3, we provide a formalization of the association rule mining task and the needed notations. Then, in Section 4, we emphasize the need for primitives that work on more or less condensed representations for the frequent itemsets, e.g., the (frequent) δ-free and closed itemsets. In Section 5, the use of these condensed representations for both the optimization of inductive queries and sophisticated rule post-processing is discussed.

2   The MINE RULE Operator [23]

Throughout this paper, we use the MINE RULE query example of Figure 1 on the relational database of Table 1. The database records several transactions made by three customers in a store on different dates. The result of such a query is a set of frequent and valid association rules. A rule like Coffee Boots $\Rightarrow$ Darts is frequent if enough customers buy within a same transaction Coffee, Boots and Darts. This rule is said valid if a customer who buys Coffee and Boots tends to buy Darts either.

Association rules are mined from a so-called *transactional database* that must be specified within the query. The FROM clause of the query specifies which part of the relational database (using any valid SQL query) is considered to construct the transactional database (e.g., given the used WHERE clause, only the transactions done after Nov. 8 are used). The GROUP BY clause specifies that the rows of the purchase table are grouped by transactions to form the rows of the transactional database (e.g., another choice would have been to group the rows by customers). In our query example, the result of this grouping/selection step is the transactional database T of Figure 2.

The specified transactional database is used to perform association rule mining under constraints. The SELECT clause specifies that the body and head of the rules are products (a rule has the form body $\Rightarrow$ head where body and head are sets of products) and that their size is greater than one (with no upper bound). This query also defines the constraints that must be fulfilled by the rules. The rules must be frequent (with a frequency threshold of 0.5), valid (with a confidence threshold of 0.7), and must satisfy the other constraints expressed in the SELECT clause: $\mathcal{C}_a(X \Rightarrow Y) \equiv \forall A \in Y, A.price > 100$ and $\mathcal{C}_b(X \Rightarrow Y) \equiv |(X \cup Y) \cap \{\text{Album}, \text{Boots}\}| \leq 1$. $\mathcal{C}_a$ means that all products in the head of the rule must have a price greater than 100 and $\mathcal{C}_b$ means that the rule must contain at most one product out of $\{\text{Album}, \text{Boots}\}$. Finally, the answer to this query is the set of rules that satisfy $\mathcal{C}_{\text{freq}} \wedge \mathcal{C}_{\text{conf}} \wedge \mathcal{C}_a \wedge \mathcal{C}_b$ on the transactional database T of Figure 2.

Let us now review some of the open problems with the MINE RULE proposal.

- *Data selection and preprocessing.* Indeed, query languages based on SQL enable to use the full power of this standard query language for data selection.

Table 1. Part of the purchase table.

tr.	cust.	product	date	price
1	$cust_1$	Coffee	Nov. 8	20
1	$cust_1$	Darts	Nov. 8	50
2	$cust_2$	Album	Nov. 9	110
2	$cust_2$	Boots	Nov. 9	120
2	$cust_2$	Coffee	Nov. 9	20
2	$cust_2$	Darts	Nov. 9	50
3	$cust_1$	Boots	Nov. 9	120
3	$cust_1$	Coffee	Nov. 9	20
4	$cust_3$	Album	Nov. 10	110
4	$cust_3$	Coffee	Nov. 10	20
⋮	⋮	⋮	⋮	⋮

```
MINE RULE result AS
SELECT DISTINCT 1..n product AS BODY,
   1..n product AS HEAD, SUPPORT, CONFIDENCE
   WHERE HEAD.price> 100 AND
   |(HEAD ∪ BODY) ∩ {Album, Boots}| ≤ 1
FROM purchase WHERE date > Nov. 8
GROUP BY transaction
EXTRACTING RULES WITH SUPPORT: 0.5, CONFIDENCE: 0.7
```

Fig. 1. A MINE RULE query on the purchase database.

It is out of the scope of this paper to discuss this phase but it is interesting to note that MINE RULE offers no specific primitive for data preprocessing (e.g., discretization) and that the other languages like MSQL offer just a few [16]. Preprocessing remains ad-hoc for many data mining processes and it is often assumed that it is performed beforehand by means of various software tools.

– The specification of the type of rules to be mined is defined in MINE RULE by the SELECT clause. It enables the definition of simple syntactic constraints, the specification of the sorts of attributes, and the definition of the so-called *mining conditions* that can make use of some background knowledge. Using MINE RULE, it is assumed that this knowledge has been encoded within the relational database.

– In MINE RULE, it is possible to define minimal frequency and minimal confidence for the desired rules.

– *Rule post-processing.* When using MINE RULE, no specific post-processing primitive is offered. This contrasts with the obvious needs for pattern post-processing in unsupervized data mining processes like association rule mining. Indeed, extracted rules can be stored under a relational schema and then be queried by means of SQL. However, it has been shown (see, e.g., [5]) that simple post-processing queries are then quite difficult to express.

To the best of our knowledge, in the MINE RULE architecture, the collection of frequent itemsets and their frequencies is not directly available for further use. It means that the computation of other interestingness measures like the J-measure [29] is not possible without looking again at the data. For rule post-processing, MSQL is richer than the other languages in its offer of few built-in post-processing primitives (it reserves a dedicated operator, SelectRules for these purposes and primitives for crossing over the rules to the data). However, none of the proposed languages supports complex post-processing processes (e.g., the computation of non redundant rules) as needed in real-life association rule mining.

It is useful to abstract the meaning of such mining queries. A simple model has been introduced in [22] that considers a data mining process as a sequence of queries over the data but also the *theory* of the data. Given a language $\mathcal{L}$ of patterns (e.g., association rules), the theory of a database $\mathbf{r}$ with respect to $\mathcal{L}$ and a selection predicate q is the set $Th(\mathbf{r}, \mathcal{L}, q) = \{\phi \in \mathcal{L} \mid q(\mathbf{r}, \phi) \text{ is true}\}$. The predicate q indicates whether a pattern ϕ is considered interesting (e.g., q specifies that ϕ is "frequent" in $\mathbf{r}$). The selection predicate can be defined as a combination (boolean expression) of atomic constraints that have to be satisfied by the patterns. Some of them refer to the "behavior" of a pattern in the data (e.g., its "frequency" in $\mathbf{r}$ is above a threshold), some others define syntactical restrictions on the desired patterns (e.g., its "length" is below a threshold). Preprocessing defines $\mathbf{r}$, the mining phase is often the computation of the specified theory while post-processing can be considered as a querying activity on a materialized theory or the computation of a new theory. A "generate and test" approach that would enumerate the sentences of $\mathcal{L}$ and then test the selection predicate q is generally impossible. A huge effort has concerned the "active" use of the primitive constraints occurring in q to have a tractable evaluation of useful mining queries.

Indeed, given the (restricted) collection of primitives offered by the MINE RULE operator, it is possible to have an efficient implementation thanks to the availability of efficient algorithms for computing frequent itemsets from huge but sparse databases [2,26]. To further extend both the efficiency of single query evaluation (especially in the difficult contexts where expressive constraints are used and/or the data are highly-correlated), the offered primitives for post-processing and the optimization of sequence of queries, we now consider an abstract framework in which the impact of the so-called *condensed representations* of frequent patterns can be emphasized.

3   Association Rule Mining Queries

Assume that Items is a finite set of symbols denoted by capital letters, e.g., Items= $\{A, B, C, \dots\}$. A *transactional database* is a collection of rows where each row is a subset of Items. An *itemset* is a subset of Items. A row r *supports* an itemset S if $S \subseteq r$. The *support* (denoted support(S)) of an itemset S is the multi-set of all rows of the database that support S. The *frequency* of an itemset S is $|\text{support}(S)|/|\text{support}(\emptyset)|$ and is denoted $\mathcal{F}(S)$. Figure 2 provides an example of a transactional database and the supports and the frequencies of some itemsets. We often use a string notation for itemsets, e.g., AB for $\{A, B\}$.

An association rule is denoted $X \Rightarrow Y$ where $X \cap Y = \emptyset$ and $X \subseteq \text{Items}$ is the *body* of the rule and $Y \subseteq \text{Items}$ is the *head* of the rule. The *support* and *frequency* of a rule are defined as the support and frequency of the itemset $X \cup Y$. A row r *supports* a rule $X \Rightarrow Y$ if it supports $X \cup Y$. A row r is an exception for a rule $X \Rightarrow Y$ if it supports X and it does not support Y. The *confidence* of the rule is $\mathcal{CF}(X \Rightarrow Y) = \mathcal{F}(X \Rightarrow Y)/\mathcal{F}(X) = \mathcal{F}(X \cup Y)/\mathcal{F}(X)$. The confidence of the rule gives the conditional probability that a row supports $X \cup Y$ when it supports X. A rule with a confidence of one has no exception and is called a *logical rule*. Frequency and confidence are two popular *evaluation functions* for association rules. Interestingly, the association rule mining task is not well specified as a theory computation. Indeed, in this kind of process, we need not only the patterns that satisfy q but also the results of some evaluation functions for each of these selected patterns.

We now define constraints on itemsets and rules.

Definition 1 (constraint). *If $\mathcal{B}$ denotes the set of all transactional databases and 2^{Items} the set of all itemsets, an itemset constraint $\mathcal{C}$ is a predicate over $2^{\text{Items}} \times \mathcal{B}$. Similarly, a rule constraint is a predicate over $\mathcal{R} \times \mathcal{B}$ where $\mathcal{R}$ is the set of association rules. An itemset $S \in 2^{\text{Items}}$ (resp. a rule R) satisfies a constraint $\mathcal{C}$ in the database $B \in \mathcal{B}$ iff $\mathcal{C}(S, B) = \text{true}$ (resp. $\mathcal{C}(R, B) = \text{true}$). When it is clear from the context, we write $\mathcal{C}(S)$ (resp. $\mathcal{C}(R)$). Given a subset I of Items, we define $\text{SAT}_{\mathcal{C}}(I) = \{S \in I, S \text{ satisfies } \mathcal{C}\}$ for an itemset constraint (resp. if J is a subset of $\mathcal{R}$, $\text{SAT}_{\mathcal{C}}(J) = \{R \in J, R \text{ satisfies } \mathcal{C}\}$ for a rule constraint). $\text{SAT}_{\mathcal{C}}$ denotes $\text{SAT}_{\mathcal{C}}(2^{\text{Items}})$ or $\text{SAT}_{\mathcal{C}}(\mathcal{R})$.*

We can now define the frequency constraint for itemsets and the frequency and confidence constraints for association rules. $\mathcal{C}_{\text{freq}}(S) \equiv \mathcal{F}(S) \geq \gamma$, $\mathcal{C}_{\text{freq}}(X \Rightarrow Y) \equiv \mathcal{F}(X \Rightarrow Y) \geq \gamma$, $\mathcal{C}_{\text{conf}}(X \Rightarrow Y) \equiv \mathcal{CF}(X \Rightarrow Y) \geq \theta$ where γ is the frequency threshold and θ the confidence threshold. A rule that satisfies $\mathcal{C}_{\text{freq}}$ is said frequent. A rule that satisfies $\mathcal{C}_{\text{conf}}$ is said valid.

Example 1 *Consider the dataset of Figure 2 where $\text{Items} = \{A, B, C, D\}$. If the frequency threshold is 0.5, then with the constraint $\mathcal{C}_b(S) \equiv |S \cap \{A, B\}| \leq 1$, $\text{SAT}_{\mathcal{C}_{\text{freq}} \wedge \mathcal{C}_b} = \{\emptyset, A, B, C, AC, BC\}$ If the confidence threshold is 0.7, then the rules satisfying the constraint of Figure 1 are $\text{SAT}_{\mathcal{C}_a \wedge \mathcal{C}_b \wedge \mathcal{C}_{\text{freq}} \wedge \mathcal{C}_{\text{conf}}} = \{\emptyset \Rightarrow A, C \Rightarrow A\}$.*

Let us now formalize that inductive queries that return itemsets or rules must also provide the results of the evaluation functions for further use.

Definition 2 (itemset query). *A itemset query is a pair $(\mathcal{C}, \mathbf{r})$ where $\mathbf{r}$ is a transactional database and $\mathcal{C}$ is an itemset constraint. The result of a query $Q = (\mathcal{C}, \mathbf{r})$ is defined as the set $Res(Q) = \{(S, \mathcal{F}(S)) | S \in 2^{\text{Items}} \wedge \mathcal{C}(S) = \text{true}\}$.*

$$T = \begin{array}{l|l} t_2 & \text{ABCD} \\ t_3 & \text{BC} \\ t_4 & \text{AC} \\ t_5 & \text{AC} \\ t_6 & \text{ABCD} \\ t_7 & \text{ABC} \end{array}$$

Itemset	Support	Frequency
A	$\{t_2, t_4, t_5, t_6, t_7\}$	0.83
B	$\{t_2, t_3, t_6, t_7\}$	0.67
AB	$\{t_2, t_6, t_7\}$	0.5
AC	$\{t_2, t_4, t_5, t_6, t_7\}$	0.83
CD	$\{t_2, t_6\}$	0.33
ACD	$\{t_2, t_6\}$	0.33

Fig. 2. Supports and frequencies of some itemsets in a transactional database. This database is constructed during the evaluation of the MINE RULE query of Figure 1 from the purchase table of Table 1

Definition 3 (association rule query). *An association rule query is a pair* $(\mathcal{C}, \mathbf{r})$ *where* $\mathbf{r}$ *is a transactional database and* $\mathcal{C}$ *is an association rule constraint. The result of a query* $Q = (\mathcal{C}, \mathbf{r})$ *is defined as the set* $Res(Q) = \{(R, \mathcal{F}(R), \mathcal{CF}(R)) \mid R \in \mathcal{R} \wedge \mathcal{C}(R) = true\}.$

The classical association rule mining problem can be stated in an association rule query where the constraint part is the conjunction of the frequency and confidence constraint [1]. Our framework enables more complex queries and does not require that the frequency and/or frequency constraints appear in $\mathcal{C}$. However, if the constraint $\mathcal{C}$ is not enough selective, the query will not be tractable. Selectivity can not been predicted beforehand. Fortunately, when the constraint has some nice properties, e.g., it is a conjunction of anti-monotone and monotone atomic constraints, efficient evaluation strategies have been identified (see the end of this section).

Our definition of an association rule query can also be modified to include other quality measures (e.g., the J-measure [29]) and not only the frequency and the confidence.

Computing the result of the classical association rule mining problem is generally done in two steps [2]: first the computation of all the frequent itemsets and their frequency and then the computation of every valid association rule that can be made from disjoint subsets of each frequent itemset. This second step is far less expensive than the first one because no access to the database is needed: only the collection of the frequent itemsets and their frequencies are needed.

To compute the result of an arbitrary association rule query, the same strategy can be used. First, derive an itemset query from the association rule query, then compute the result of this query using the transactional database and finally generate the association rules from the itemsets. For the first step, there is no general method. This is generally done in an ad-hoc manner (see Example 2) and supporting this remains an open problem. The generation of the rules can be performed by the following algorithm:

Algorithm 1 (Rule_Gen) *Given an association rule query* $(\mathbf{r}, \mathcal{C})$ *and the result Res of the itemset query, do:*

For each pair $(S, \mathcal{F}(S)) \in Res$ and for each $T \subset S$
   Construct the rule $R := T \Rightarrow (S - T)$
   Compute $\mathcal{F}(R) := \mathcal{F}(S)$ and $\mathcal{CF}(R) := \mathcal{F}(S)/\mathcal{F}(T)$.
   Output $(R, \mathcal{F}(R), \mathcal{CF}(R))$ if it satisfies the rule constraint C.

Since the database is used only during the computation of itemsets, the generation of rules is efficient.

Example 2 *The constraint used in the query of Figure 1 is: $C_{ar}(X \Rightarrow Y) = C_{\text{freq}} \wedge C_{\text{conf}} \wedge C_a(X \Rightarrow Y) \wedge C_b(X \Rightarrow Y)$ where $C_a(X \Rightarrow Y) \equiv \forall A \in Y, A.price > 100$ and $C_b(X \Rightarrow Y) \equiv |(X \cup Y) \cap \{\text{Album}, \text{Boots}\}| \leq 1$. C_b can be rewritten as an itemset constraint: $C_b(S) \equiv |S \cap \{\text{Album}, \text{Boots}\}| \leq 1$. Furthermore, since (as specified in the MINE RULE query) rules cannot have an empty head, $C_a(X \Rightarrow Y) \equiv \forall A \in Y, A.price > 100 \wedge C'_a(X \cup Y)$ where $C'_a(S) \equiv |S \cap \{I \in \text{Items}, I.price > 100\}| \geq 1$ is a useful itemset constraint.*

Finally, we can derive an itemset query $Q_i = (C_i, \mathbf{r})$ with the constraint $C_i = C_{\text{freq}} \wedge C_b \wedge C'_a$ and be sure that the result of this itemset query will allow the generation of the result of the association rule query $Q = (C_{ar}, \mathbf{r})$ using Algorithm 1.

The efficiency of the extraction of the answer to the itemset query relies on the possibility to use constraints during the itemset computation. A classical result is that effective safe pruning can be achieved when considering anti-monotone constraints [22,26]. It relies on the fact that if an itemset violates an anti-monotone constraint then all its supersets violate it as well and therefore this itemset and its supersets can be pruned.

Definition 4 (Anti-monotonicity). *An anti-monotone constraint is a constraint C such that for all itemsets S, S': $(S' \subseteq S \wedge C(S)) \Rightarrow C(S')$.*

The prototypical anti-monotone constraint is the frequency constraint. The constraint C_b of Example 2 is another anti-monotone constraint and many other examples can be found, e.g., in [26]. Notice that the conjunction or disjunction of anti-monotone constraints is anti-monotone.

The monotone constraints can also be used to improve the efficiency of itemset extraction (optimization of the candidate generation phase that prevents to consider candidates that do not satisfy the monotone constraint) [17]. However, pushing monotone constraints sometimes increases the computation times since it prevents effective pruning based on anti-monotone constraints [30,9,12].

Definition 5 (Monotonicity). *A monotone constraint is a constraint C such that for all itemsets S, S': $(S \subseteq S' \wedge S$ satisfies $C) \Rightarrow S'$ satisfies C.*

Example 3 C'_a *(see Example 2), $C(S) \equiv \{\text{A}, \text{B}, \text{C}, \text{D}\} \subset S$, $C(S) \equiv$ Sum$(S.price) > 100$ (the sum of the prices of items from S is greater than 100) and $C(S) \equiv S \cap \{\text{A}, \text{B}, \text{C}\} \neq \emptyset$ are examples of monotone constraints.*

4 Condensed Representations of Frequent Sets

To answer an association rule query, we must be able to provide efficiently the
frequency of many itemsets (see Algorithm 1). Computing the frequent itemsets
is a first solution. Another one is to use condensed representations with respect to
frequency queries. Condensed representation is a general concept (see, e.g., [21]).
In our context, the intuition is to substitute to the database or the collection
of the frequent itemsets, another representation from which we can derive the
whole collection of the frequent itemsets and their frequencies. In this paper,
given a set S of pairs $(X, \mathcal{F}(X))$, we are interested in condensed representations
of S that are subsets of S with two properties: (1) It is much smaller than S
and faster to compute, and (2), the whole set S can be generated from the
condensed representation with no access to the database, i.e., efficiently. User-
defined constraints can also be used to further optimize the computation of
condensed representations [17].

Several algorithms exist to compute various condensed representations of frequent itemsets: CLOSE [27], CLOSET [28], CHARM [31], MIN-EX [6], or PASCAL
[4]. These algorithms compute different condensed representations: the frequent
closed itemsets (CLOSE, CLOSET, CHARM), the frequent free itemsets (MIN-EX,
PASCAL), or the frequent δ-free itemsets for MIN-EX. Also, a new promising
condensed representation, the disjoint-free itemsets, has been proposed in [11].
These algorithms enable tractable extractions from dense and highly-correlated
data, i.e., extractions for frequency thresholds on which APRIORI-like algorithms
are intractable. Let us now discuss two representations on which we have been
working: the closed itemsets and the δ-free itemsets.

Definition 6 (closures and closed itemsets). *The closure of an itemset S
(denoted by* closure(S)*) is the maximal (for set inclusion) superset of S which
has the same support than S. A closed itemset is an itemset that is equal to its
closure.*

The next proposition shows how to compute the frequency of an itemset
using the collection of the frequent closed itemsets efficiently, i.e., with no access
to the database.

Proposition 1 *Given an itemset S and the set of frequent closed itemsets,*

- *If S is not included in a frequent closed itemset then S is not frequent.*
- *Else S is frequent and $\mathcal{F}(S) = \text{Max}\{\mathcal{F}(X), \ S \subseteq X \text{ and } X \text{ is closed}\}$.*

Using this proposition, it is possible to design an algorithm to compute the
result of a frequent itemset query using the frequent closed itemsets. This algorithm is not given here (see, e.g., [27,4]). As a result, γ-frequent closed itemsets are like the γ-frequent itemsets a $\gamma/2$-adequate representation for frequency
queries [6], i.e., the error on the exact frequency for any itemset is bound by $\gamma/2$
(the $\gamma/2$ value is given to infrequent itemsets and the frequency of any frequent
itemset is known exactly).

Example 4 *In the transactional database of Figure 2, if the frequency threshold is 0.2, every itemset is frequent (16 frequent itemsets). The frequent closed sets are* C, AC, BC, ABC, *and* ABCD *and we can use the previous property to get the frequency of non-closed itemsets from closed ones (e.g.,* $\mathcal{F}(AB) = \mathcal{F}(ABC)$ *since* ABC *is the smallest closed superset of* AB*).*

We can compute the closed sets from the free sets.

Definition 7 (free itemset). *An itemset S is free if no logical rule holds between its items, i.e., it does not exist two distinct itemsets X, Y such that* $S = X \cup Y$, $Y \neq \emptyset$ *and* $X \Rightarrow Y$ *is a logical rule.*

Example 5 *In the transactional database of Figure 2, if the frequency threshold is 0.2, the frequent free sets are* $\emptyset$, A, B, D, *and* AB.

The closed sets are the closure of the free one. Freeness is an anti-monotone property and thus can be used efficiently, e.g., within a level-wise algorithm.

When they can be computed, closed itemsets constitute a good condensed representation (see, e.g., [6] for experiments with real-life dense and correlated data sets). The free sets can be generalized to δ-free itemsets[2]. Representations based on δ-free itemsets are quite interesting when it is not possible to mine the closed sets, i.e., when the computation is intractable given the user-defined frequency threshold. Indeed, algorithms like CLOSE [27] or PASCAL [4] use logical rule to prune candidate itemsets because their frequencies can be inferred from the frequencies of free/closed itemsets. However, to be efficient, these algorithms need that such logical rules hold in the data. If it is not the case, then the frequent free sets are exactly the frequent itemsets and we get no improvement over APRIORI-like algorithms.

The MIN-EX algorithm introduced in [6,8] computes δ-free itemsets. The concept of closure is extended, providing new possibilities for pruning. However, we must trade this efficiency improvement against precision: the frequency of the frequent itemsets are only known within a bounded error. The MIN-EX algorithm uses rules with few exceptions to further prune the search space. Given an itemset $S = X \cup Y$ and a rule $Y \Rightarrow Z$ with less than δ exceptions, then the frequency of $X \cup Y \cup Z$ can be approximated by the frequency of S. More formally, MIN-EX uses an extended notion of closure.

Definition 8 (δ-closure and δ-free itemsets). *Let δ be an integer and S an itemset. The δ-closure of S,* $\text{closure}_\delta(S)$ *is the maximal (w.r.t. the set inclusion) superset Y of S such that for every item $A \in Y - S$, $|\text{Support}(S \cup \{A\})|$ is at least $|\text{Support}(S)| - \delta$. An itemset S is δ-free if no association rule with less than δ exceptions holds between its subsets.*

Example 6 *In the transactional database of Figure 2, if the frequency threshold is 0.2 and $\delta = 1$, the frequent 1-free sets are* $\emptyset$, A, B, *and* D.

---

[2] There is no such generalization for closed sets

Notice that with $\delta = 0$, it is the same closure operator than for CLOSE, i.e., closure$_0$ = closure. Larger values of δ leads to more efficient pruning (there are less δ-free itemsets) but also larger errors on the frequencies of itemsets when they are regenerated from the δ-free ones (see below).

The output of the MIN-EX algorithm is formally given by the three following sets: $FF(\mathbf{r}, \gamma, \delta)$ is the set of the γ-frequent δ-free itemsets, $IF(\mathbf{r}, \gamma, \delta)$ is the set of the minimal (w.r.t. the set inclusion) infrequent δ-free itemsets (i.e., the infrequent δ-free itemsets whose all subsets are γ-frequent). $FN(\mathbf{r}, \gamma, \delta)$ is the set of the minimal γ-frequent non-δ-free itemsets (i.e., the γ-frequent non-δ-free itemsets whose all subsets are δ-free). The two pairs (FF, IF) and (FF, FN) are two condensed representations based on δ-free itemsets. The next proposition shows that it is possible to compute an approximation of the frequency of an itemset using one of these two condensed representations.

Proposition 2 *Let S be an itemset. If there exists $X \in IF(\mathbf{r}, \gamma, \delta)$ such that $X \subseteq S$ then S is infrequent. If $S \notin FF(\mathbf{r}, \gamma, \delta)$ and there does not exist $X \in FN(\mathbf{r}, \gamma, \delta)$ such that $X \subseteq S$ then S is infrequent. In these two cases, the frequency of S can be approximated by $\gamma/2$ Else, let F be the δ-free itemset such that: $\mathcal{F}(F) = \text{Min}\{\mathcal{F}(X),\ X \subseteq S$ and X is δ-free$\}$. Assuming that $n_S = |\text{support}(S)|$ and $n_F = |\text{support}(F)|$, then $n_F \geq n_S \geq n_F - \delta(|S| - |F|)$, or, dividing this by n, the number of rows in $\mathbf{r}$, $\mathcal{F}(F) \geq \mathcal{F}(S) \geq \mathcal{F}(F) - \frac{\delta}{n}(|S| - |F|)$.*

Typical δ values range from zero to a few hundreds. With a database size of several tens of thousands of rows, the error made is below few percents [8].

Using Proposition 2, it is also possible to regenerate an approximation of the answer to a frequent itemset query from one of the condensed representation (FF, IF) or (FF, FN):

- The frequency of an itemset is approximated with an error bound given by Proposition 2 (notice that this error is computed during the regeneration and thus can be presented to the user with the frequency of each itemset).
- Some of the computed itemsets might be infrequent because the uncertainty on their frequencies does not enable to classify them as frequent or infrequent (e.g., if $\gamma = 0.5$ and the $\mathcal{F}(X) = 0.49$ with an error of 0.02).

If $\delta = 0$, then the two condensed representations enable to regenerate exactly the answer to a frequent itemset query.

Given an arbitrary itemset query $Q = (\mathcal{C}, \mathbf{r})$, there are therefore two solutions to compute its answer:

- Pushing the anti-monotone and monotone part of the constraint $\mathcal{C}$ as sketched in Section 3.
- Using condensed representation to answer a more general query (with only the frequency constraint) and then filter the itemsets that do not verify the constraint $\mathcal{C}$.

We now consider how to combine these two methods for mining condensed representations that satisfy a conjunction of an anti-monotone and a monotone constraint. In [17], we presented an algorithm to perform this extraction. This algorithm uses an extension of the δ-free itemsets, the *contextual δ-free itemsets*.

Definition 9 (contextual δ-free itemset). *An itemset S is contextual δ-free with respect to a monotone constraint C_m if it does not exist two distinct subsets X, Y of S such that X satisfies C_m and $X \Rightarrow Y$ has less than δ exceptions.*

The input and output of this algorithm are formalized as follows:

Input: a query $Q = (C_{am} \wedge C_m, \mathbf{r})$ where C_{am} is an anti-monotone constraint and C_m a monotone constraint.
Output: three collections FF, IF, FN and, if $\delta = 0$, the collection $\mathcal{O}$.

- $FF = \{(S, \mathcal{F}(S)) | S$ is contextual δ-free and $C_{am}(S) \wedge C_m(S)$ is true$\}$,
- IF and FN are defined as for MIN-EX and
- $\mathcal{O} = \{(\texttt{closure}(S), \mathcal{F}(S)) | S$ is free and $C_{\text{freq}}(S) \wedge C_{am}(S)$ is true$\}$.

These collections give three condensed representations $\mathcal{O}$ (if $\delta = 0$), (FF, IF) and (FF, FN). The regeneration of the answer to the query Q using the collection $\mathcal{O}$ of closed itemsets can be done by:

Given an itemset S
If $C_m(S)$ is true, then use Proposition 1 to compute $\mathcal{F}(S)$
If $C_{am}(S)$ is true then output $(S, \mathcal{F}(S))$.

When considering (FF, IF) or (FF, FN):

If $C_m(S)$ is true, then use (FF, IF) or (FF, FN) as in Proposition 2 to compute $\mathcal{F}(S)$.
If $C_{am}(S)$ is true or unknown then output $(S, \mathcal{F}(S))$

The result of $C_{am}(S)$ can be unknown due to the uncertainty on the frequency (if $\delta \neq 0$).

5 Uses of Condensed Representations

Let us now sketch several applications of such condensed representations.

Optimization of MINE RULE *queries.* It is clear that the given condensed representations of the frequent patterns can be used, in a transparent way for the end-user, for optimization purposes. In such a context, we just have to replace the algorithmic core that concerns frequent itemset mining by our algorithms that compute free/closed itemsets and then derive the whole collection of the frequent itemsets. Also, the optimized way to push conjunction of monotone and anti-monotone constraints might be implemented.

Condensed representations have other interesting applications beyond the optimization of an association rule mining query.

Generation of non-redundant association rules. One of the problems in association rule mining from real-life data is the huge number of extracted rules. However, many of the rules are in some sense redundant and might be useless, e.g., AB $\Rightarrow$ C is not interesting if A $\Rightarrow$ BC has the same confidence. In [3], an algorithm is presented to extract a minimal cover of the set of frequent association rules. This set is generated from the closed and free itemsets. This cover can be generated by considering only rules of the form $X \Rightarrow (Y - X)$ where X is a free itemset and Y is a closed itemset containing X. It leads to a much smaller collection of association rules than the one computed from itemsets using Algorithm 1. In this volume, [19] considers other concise representations of association rules. In our practice, post-processing the discovered rules can clearly make use of the properties of the free and close sets. In other terms, materializing these collections can be useful for post-processing, not only the optimization of the mining phase. For instance, it makes sense to look for association rules that contain free itemsets as their left-hand sides and some targeted attributes on the right-hand sides without any minimal frequency constraint. It might remain tractable, thanks to the anti-monotonicity of freeness (extraction of the whole collection of the free itemsets), and need a reasonable amount of computation when computing the frequency and the confidence of each candidate rule.

Using condensed representation as a knowledge cache. A user generally submits a query, gets the results and refines it until he is satisfied by the extracted patterns. Since computing the result for one single query can be expensive (several minutes up to several hours), it is highly desirable that the data mining system optimizes sequences of queries. A classical solution is to cache the results of previous queries to answer faster to new queries. This has been studied by caching itemsets (e.g., [13,25]). Most of these works require that some strong relation holds between the queries like inclusion or equivalence. Caching condensed representations seems quite natural and we began to study the use of free itemsets for that purpose [18]. In [18], we assume the user defines constraints on closed sets and can refine them in a sequence of queries. Free sets from previous queries are put in a cache. A cache of free itemsets is much smaller than a cache containing itemsets and our algorithm ensures that the intersection between the union of the results of all previous queries and the result of the new query is not recomputed. Finally, we do not make any assumption on the relation between two queries in the sequence. The algorithm improves the performance of the extraction with respect to an algorithm that mines the closed sets without making use of the previous computations. The speedup is roughly equal to the relative size of the intersection between the answer to a new query and the content of the cache. Again, such an optimization could be integrated into the MINE RULE architecture in a transparent way.

6   Conclusion

Even though this paper has emphasized the use of frequent itemsets for association rule mining, the interest of inductive querying on itemsets goes far beyond this popular mining task. For instance, constrained itemsets and their frequencies can be used for computing similarity measures between attributes and thus for clustering tasks (see, e.g., [24]). It can also be used for the discovery of more general kinds of rules, like rules with restricted forms of disjunctions or negations [21,7] and the approximation of the joint distribution [20]. Our future line of work will be, (1) to investigate the multiple uses of the condensed representations of frequent itemsets, and (2) to study evaluation strategies for association rule mining queries when we have complex selection criteria (i.e., general boolean expression instead of conjunctions of monotone and anti-monotone constraints).

Acknowledgments. The authors thank the researchers from the CINQ consortium for interesting discussions and more particularly, Luc De Raedt, Mika Klemettinen, Heikki Mannila, Rosa Meo, and Christophe Rigotti.

References

1. R. Agrawal, T. Imielinski, and A. Swami. Mining association rules between sets of items in large databases. In *Proceedings SIGMOD'93*, pages 207–216, Washington, USA, 1993. ACM Press.
2. R. Agrawal, H. Mannila, R. Srikant, H. Toivonen, and A. I. Verkamo. Fast discovery of association rules. In *Advances in Knowledge Discovery and Data Mining*, pages 307–328. AAAI Press, 1996.
3. Y. Bastide, N. Pasquier, R. Taouil, G. Stumme, and L. Lakhal. Mining minimal non-redundant association rules using frequent closed itemsets. In *Proceedings CL 2000*, volume 1861 of *LNCS*, pages 972–986, London, UK, 2000. Springer-Verlag.
4. Y. Bastide, R. Taouil, N. Pasquier, G. Stumme, and L. Lakhal. Mining frequent patterns with counting inference. *SIGKDD Explorations*, 2(2):66–75, Dec. 2000.
5. M. Botta, J.-F. Boulicaut, C. Masson, and R. Meo. A comparison between query languages for the extraction of association rules. In *Proceedings DaWaK'02*, Aix en Provence, F, 2002. Springer-Verlag. To appear.
6. J.-F. Boulicaut and A. Bykowski. Frequent closures as a concise representation for binary data mining. In *Proceedings PAKDD'00*, volume 1805 of *LNAI*, pages 62–73, Kyoto, JP, 2000. Springer-Verlag.
7. J.-F. Boulicaut, A. Bykowski, and B. Jeudy. Towards the tractable discovery of association rules with negations. In *Proceedings FQAS'00*, Advances in Soft Computing series, pages 425–434, Warsaw, PL, Oct. 2000. Springer-Verlag.
8. J.-F. Boulicaut, A. Bykowski, and C. Rigotti. Approximation of frequency queries by means of free-sets. In *Proceedings PKDD'00*, volume 1910 of *LNAI*, pages 75–85, Lyon, F, 2000. Springer-Verlag.
9. J.-F. Boulicaut and B. Jeudy. Using constraint for itemset mining: should we prune or not? In *Proceedings BDA'00*, pages 221–237, Blois, F, 2000.
10. J.-F. Boulicaut, M. Klemettinen, and H. Mannila. Modeling KDD processes within the inductive database framework. In *Proceedings DaWaK'99*, volume 1676 of *LNCS*, pages 293–302, Florence, I, 1999. Springer-Verlag.
11. A. Bykowski and C. Rigotti. A condensed representation to find frequent patterns. In *Proceedings PODS'01*, pages 267–273, Santa Barbara, USA, 2001. ACM Press.

12. M. M. Garofalakis, R. Rastogi, and K. Shim. SPIRIT: Sequential pattern mining with regular expression constraints. In *Proceedings VLDB'99*, pages 223–234, Edinburgh, UK, 1999. Morgan Kaufmann.
13. B. Goethals and J. van den Bussche. On implementing interactive association rule mining. In *Proceedings of the ACM SIGMOD Workshop DMKD'99*, Philadelphia, USA, 1999.
14. J. Han and M. Kamber. *Data Mining: Concepts and techniques*. Morgan Kaufmann Publishers, San Francisco, USA, 2000. 533 pages.
15. T. Imielinski and H. Mannila. A database perspective on knowledge discovery. *Communications of the ACM*, 39(11):58–64, 1996.
16. T. Imielinski and A. Virmani. MSQL: A query language for database mining. *Data Mining and Knowledge Discovery*, 3(4):373–408, 1999.
17. B. Jeudy and J.-F. Boulicaut. Optimization of association rule mining queries. *Intelligent Data Analysis*, 6(5), 2002. To appear.
18. B. Jeudy and J.-F. Boulicaut. Using condensed representations for interactive association rule mining. In *Proceedings PKDD'02*, Helsinki, FIN, 2002. Springer-Verlag. To appear.
19. M. Kryszkiewicz. Concise representations of association rules. In *Proceedings of the ESF Exploratory Workshop on Pattern Detection and Discovery*, London, UK, 2002. Springer-Verlag. To appear in this volume.
20. H. Mannila and P. Smyth. Approximate query answering with frequent sets and maximum entropy. In *Proceedings ICDE'00*, page 309, San Diego, USA, 2000. IEEE Computer Press.
21. H. Mannila and H. Toivonen. Multiple uses of frequent sets and condensed representations. In *Proceedings KDD'96*, pages 189–194, Portland, USA, 1996. AAAI Press.
22. H. Mannila and H. Toivonen. Levelwise search and borders of theories in knowledge discovery. *Data Mining and Knowledge Discovery*, 1(3):241–258, 1997.
23. R. Meo, G. Psaila, and S. Ceri. An extension to SQL for mining association rules. *Data Mining and Knowledge Discovery*, 2(2):195–224, 1998.
24. P. Moen. *Attribute, Event Sequence, and Event Type Similarity Notions for Data Mining*. PhD thesis, Department of Computer Science, P.O. Box 26, FIN-00014 University of Helsinki, Jan. 2000.
25. B. Nag, P. M. Deshpande, and D. J. DeWitt. Using a knowledge cache for interactive discovery of association rules. In *Proceedings SIGKDD'99*, pages 244–253. ACM Press, 1999.
26. R. Ng, L. V. Lakshmanan, J. Han, and A. Pang. Exploratory mining and pruning optimizations of constrained associations rules. In *Proceedings SIGMOD'98*, pages 13–24, Seattle, USA, 1998. ACM Press.
27. N. Pasquier, Y. Bastide, R. Taouil, and L. Lakhal. Efficient mining of association rules using closed itemset lattices. *Information Systems*, 24(1):25–46, 1999.
28. J. Pei, J. Han, and R. Mao. CLOSET an efficient algorithm for mining frequent closed itemsets. In *Proceedings of the ACM SIGMOD Workshop DMKD'00*, pages 21–30, Dallas, USA, 2000.
29. P. Smyth and R. M. Goodman. An information theoretic approach to rule induction from databases. *IEEE Transactions on Knowledge and Data Engineering*, 4(4):301–316, 1992.
30. R. Srikant, Q. Vu, and R. Agrawal. Mining association rules with item constraints. In *Proceedings KDD'97*, pages 67–73, Newport Beach, USA, 1997. AAAI Press.
31. M. J. Zaki. Generating non-redundant association rules. In *Proceedings ACM SIGKDD'00*, pages 34–43, Boston, USA, 2000. ACM Press.

Relational Association Rules: Getting WARMeR

Bart Goethals and Jan Van den Bussche

University of Limburg, Belgium

Abstract. In recent years, the problem of association rule mining in transactional data has been well studied. We propose to extend the discovery of classical association rules to the discovery of association rules of conjunctive queries in arbitrary relational data, inspired by the WARMR algorithm, developed by Dehaspe and Toivonen, that discovers association rules over a limited set of conjunctive queries. Conjunctive query evaluation in relational databases is well understood, but still poses some great challenges when approached from a discovery viewpoint in which patterns are generated and evaluated with respect to some well defined search space and pruning operators.

1 Introduction

In recent years, the problem of mining association rules over frequent itemsets in transactional data [9] has been well studied and resulted in several algorithms that can find association rules within a limited amount of time. Also more complex patterns have been considered such as trees [17], graphs [11,10], or arbitrary relational structures [5,6]. However, the presented algorithms only work on databases consisting of a set of transactions. For example, in the tree case [17], every transaction in the database is a separate tree, and the presented algorithm tries to find all frequent subtrees occurring within all such transactions. Nevertheless, many relational databases are not suited to be converted into a transactional format and even if this were possible, a lot of information implicitly encoded in the relational model would be lost after conversion. Towards the discovery of association rules in arbitrary relational databases, Deshaspe and Toivonen developed an inductive logic programming algorithm, WARMR [5, 6], that discovers association rules over a limited set of conjunctive queries on transactional relational databases in which every transaction consists of a small relational database itself. In this paper, we propose to extend their framework to a broader range of conjunctive queries on arbitrary relational databases.

Conjunctive query evaluation in relational databases is well understood, but still poses some great challenges when approached from a discovery viewpoint in which patterns are generated and evaluated with respect to some well defined search space and pruning operators. We describe the problems occurring in this mining problem and present an algorithm that uses a similar two-phase architecture as the standard association rule mining algorithm over frequent itemsets (Apriori) [1], which is also used in the WARMR algorithm. In the first phase, all frequent patterns are generated, but now, a pattern is a conjunctive query and

D.J. Hand et al. (Eds.): Pattern Detection and Discovery, LNAI 2447, pp. 125–139, 2002.

its support equals the number of distinct tuples in the answer of the query. The second phase generates all association rules over these patterns. Both phases are based on the general levelwise pattern mining algorithm as described by Mannila and Toivonen [12].

In Section 2, we formally state the problem we try to solve. In Section 3, we describe the general approach that is used for a large family of data mining problems. In Section 4, we describe the WARMR algorithm which is also based on this general approach. In Section 5, we describe our approach as an generalization of the WARMR algorithm and identify the algorithmic challenges that need to be conquered. In Section 6, we show a sample run of the presented approach. We conclude the paper in Section 7 with a brief discussion and future work.

2   Problem Statement

The relational data model is based on the idea of representing data in tabular form. The *schema* of a relational database describes the names of the tables and their respective sets of column names, also called attributes. The actual content of a database, is called an *instance* for that schema. In order to retrieve data from the database, several query languages have been developed, of which SQL is the standard adopted by most database management system vendors. Nevertheless, an important and well-studied subset of SQL, is the family of conjunctive queries.

As already mentioned in the Introduction, current algorithms for the discovery of patterns and rules mainly focused on transactional databases. In practice, these algorithms use several specialized data structures and indexing schemes to efficiently find their specific type of patterns, i.e., itemsets, trees, graphs, and many others. As an appropriate generalization of these kinds of patterns, we propose a framework for arbitrary relational databases in which *a pattern is a conjunctive query*.

Assume we are given a relational database consisting of a schema $\mathbf{R}$ and an instance $\mathbf{I}$ of $\mathbf{R}$. An *atomic formula* over $\mathbf{R}$ is an expression of the form $R(\bar{x})$, where R is a relation name in $\mathbf{R}$ and $\bar{x}$ is a k-tuple of variables and constants, with k the arity of R.

Definition 1. *A conjunctive query Q over $\mathbf{R}$ consists of a head and a body. The body is a finite set of atomic formulas over $\mathbf{R}$. The head is a tuple of variables occurring in the body.*

A *valuation* on Q is a function f that assigns a constant to every variable in the query. A valuation is a *matching* of Q in $\mathbf{I}$, if for every $R(\bar{x})$ in the body of Q, the tuple $f(\bar{x})$ is in $\mathbf{I}(R)$. The *answer* of Q on $\mathbf{I}$ is the set

$$Q(\mathbf{I}) := \{f(\bar{y}) \mid \bar{y} \text{ is the head of } Q \text{ and } f \text{ is a matching of } Q \text{ on } \mathbf{I}\}.$$

We will write conjunctive queries using the commonly used Prolog notation. For example, consider the following query on a beer drinkers database:

$$Q(x) \; :- \; likes(x, \text{'Duvel'}), likes(x, \text{'Trappist'}).$$

The answer of this query consists of all drinkers that like Duvel and also like Trappist.

For two conjunctive queries Q_1 and Q_2 over $\mathbf{R}$, we write $Q_1 \subseteq Q_2$ if for every possible instance $\mathbf{I}$ of $\mathbf{R}$, $Q_1(\mathbf{I}) \subseteq Q_2(\mathbf{I})$ and say that Q_1 is *contained* in Q_2. Q_1 and Q_2 are called *equivalent* if and only if $Q_1 \subseteq Q_2$ and $Q_2 \subseteq Q_1$. Note that the question whether a conjunctive query is contained in another conjunctive query is decidable [16].

Definition 2. *The* support *of a conjunctive query Q in an instance $\mathbf{I}$ is the number of distinct tuples in the answer of Q on $\mathbf{I}$. A query is called* frequent *in $\mathbf{I}$ if its support exceeds a given minimal support threshold.*

Definition 3. *An association rule is of the form $Q_1 \Rightarrow Q_2$, such that Q_1 and Q_2 are both conjunctive queries and $Q_2 \subseteq Q_1$. An association rule is called* frequent *in $\mathbf{I}$ if Q_2 is frequent in $\mathbf{I}$ and it is called* confident *if the support of Q_2 divided by the support of Q_1 exceeds a given minimal confidence threshold.*

Example 1. Consider the following two queries:

$$Q_1(x, y) :\text{-- } likes(x, \text{`Duvel'}), visits(x, y).$$
$$Q_2(x, y) :\text{-- } likes(x, \text{`Duvel'}), visits(x, y), serves(y, \text{`Duvel'}).$$

The rule $Q_1 \Rightarrow Q_2$ should then be read as follows: if a person x that likes Duvel visits bar y, then bar y serves Duvel.

A natural question to ask is why we should only consider rules over queries that are contained for any possible instance. For example, assume we have the following two queries:

$$Q_1(y) :\text{-- } likes(x, \text{`Duvel'}), visits(x, y).$$
$$Q_2(y) :\text{-- } serves(y, \text{`Duvel'}).$$

Obviously, Q_2 is not contained in Q_1 and vice versa. Nevertheless, it is still possible that for a given instance $\mathbf{I}$, we have $Q_2(\mathbf{I}) \subseteq Q_1(\mathbf{I})$, and hence this could make an interesting association rule $Q_1 \Rightarrow Q_2$, which should be read as follows: if bar y has a visitor that likes Duvel, then bar y also serves Duvel.

Proposition 1. *Every association rule $Q_1 \Rightarrow Q_2$, such that $Q_2(\mathbf{I}) \subseteq Q_1(\mathbf{I})$, can be expressed by an association rule $Q_1 \Rightarrow Q_2'$, with $Q_2' = Q_2 \cap Q_1$, and essentially has the same meaning.*

In this case the correct rule would be $Q_1 \Rightarrow Q_2$, with

$$Q_1(y) :\text{-- } likes(x, \text{`Duvel'}), visits(x, y).$$
$$Q_2(y) :\text{-- } likes(x, \text{`Duvel'}), visits(x, y), serves(y, \text{`Duvel'}).$$

Note the resemblance with the queries used in Example 1. The bodies of the queries are the same, but now we have another head. Evidently, different heads result in a different meaning of the corresponding association rule which can still be interesting. As another example, note the difference with the following two queries:

$$Q_1(x) \; :\!\!- \; likes(x, \text{'Duvel'}), visits(x, y).$$

$$Q_2(x) \; :\!\!- \; likes(x, \text{'Duvel'}), visits(x, y), serves(y, \text{'Duvel'}).$$

The rule $Q_1 \Rightarrow Q_2$ should then be read as follows: if a person x that likes Duvel visits a bar, then x also visits a bar that serves Duvel.

The goal is now to find all frequent and confident association rules in the given database.

3   General Approach

As already mentioned in the introduction, most association rule mining algorithms use the common two-phase architecture. Phase 1 generates all frequent patterns, and phase 2 generates all frequent and confident association rules.

The algorithms used in both phases are based on the general levelwise pattern mining algorithm as described by Mannila and Toivonen [12]. Given a database $\mathcal{D}$, a class of patterns $\mathcal{L}$, and a selection predicate q, the algorithm finds the "theory" of $\mathcal{D}$ with respect to $\mathcal{L}$ and q, i.e., the set $\mathit{Th}(\mathcal{L}, \mathcal{D}, q) := \{\phi \in \mathcal{L} \mid q(\mathcal{D}, \phi) \text{ is true}\}$. The selection predicate q is used for evaluating whether a pattern $Q \in \mathcal{L}$ defines a (potentially) interesting pattern in $\mathcal{D}$. The main problem this algorithm tries to tackle is to minimize the number of patterns that need to be evaluated by q, since it is assumed this evaluation is the most costly operation of such mining algorithms. The algorithm is based on a breadth-first search in the search space spanned by a specialization relation which is a partial order $\preceq$ on the patterns in $\mathcal{L}$. We say that ϕ is *more specific* than ψ, or ψ is *more general* than ϕ, if $\phi \preceq \psi$. The relation $\preceq$ is a *monotone specialization relation* with respect to q, if the selection predicate q is monotone with respect to $\preceq$, i.e., for all $\mathcal{D}$ and ϕ, we have the following: if $q(\mathcal{D}, \phi)$ and $\phi \preceq \gamma$, then $q(\mathcal{D}, \gamma)$. In what follows, we assume that $\preceq$ is a monotone specialization relation. We write $\phi \prec \psi$ if $\phi \preceq \psi$ and not $\psi \preceq \phi$. The algorithm works iteratively, alternating between *candidate generation* and *candidate evaluation*, as follows.

$C_1 := \{\phi \in \mathcal{L} \mid \text{there is no } \gamma \text{ in } \mathcal{L} \text{ such that } \phi \prec \gamma\};$
$i := 1;$
while $C_i \neq \emptyset$ **do**
    // Candidate evaluation
    $\mathcal{F}_i := \{\phi \in C_i \mid q(\mathcal{D}, \phi)\};$
    // Candidate generation
    $C_{i+1} := \{\phi \in \mathcal{L} \mid \text{for all } \gamma, \text{ such that } \phi \prec \gamma, \text{ we have } \gamma \in \bigcup_{j \leq i} \mathcal{F}_j\} \backslash \bigcup_{j \leq i} C_j;$
    $i := i + 1$
end while
return $\bigcup_{j < i} \mathcal{F}_j;$

In the generation step of iteration i, a collection C_{i+1} of new *candidate patterns* is generated, using the information available from the more general patterns in $\bigcup_{j \leq i} \mathcal{F}_j$, which have already been evaluated. Then, the selection predicate is evaluated on these candidate patterns. The collection $\mathcal{F}_{i+1}$ will consist of those patterns in C_{i+1} that satisfy the selection predicate q. The algorithm starts by constructing C_1 to contain all most general patterns. The iteration stops when no more potentially interesting patterns can be found with respect to the selection predicate.

In general, given a language $\mathcal{L}$ from which patterns are chosen, a selection predicate q and a monotone specialization relation $\preceq$ with respect to q, this algorithm poses several challenges.

1. An initial set C_1 of most general candidate patterns needs to be identified, which is not always possible for infinite languages, and hence other, maybe less optimal solutions could be required.
2. Given all patterns $\bigcup_{j \leq i} \mathcal{F}_j$ that satisfy the selection predicate up to a certain level i, the set C_{i+1} of all candidate patterns must be generated efficiently. It might be impossible to generate all but only those elements in C_{i+1}, but instead, it might be necessary to generate a superset of C_{i+1} after which the non candidate patterns must be identified and removed. Even if this identification is efficient, naively generating all possible patterns could still become infeasible if this number of patterns becomes too large. Hence, this poses two additional challenges:
   a) efficiently generate the smallest possible superset of C_{i+1}, and
   b) identify and remove each generated pattern that is no candidate pattern by efficiently checking whether all of its generalizations are in $\bigcup_{j \leq i} \mathcal{F}_j$.
3. Extract all patterns from C_{i+1} that satisfy the selection predicate q, by efficiently evaluating q on all elements in C_{i+1}.

In the next section, we identify these challenges for both phases of the association rule mining problem within the framework proposed by Dehaspe and Toivonen, and describe their solutions as implemented within the WARMR algorithm.

4   The WARMR Algorithm

As already mentioned in the introduction, a first approach towards the goal of discovering all frequent and confident association rules in arbitrary relational databases, has been presented by Dehaspe and Toivonen, in the form of an inductive logic programming algorithm, WARMR [5,6], that discovers association rules over a limited set of conjunctive queries.

4.1   Phase 1

The procedure to generate all frequent conjunctive queries is primarily based on a *declarative language bias* to constrain the search space to a subset of all conjunctive queries, which is an extensively studied subfield in ILP [13].

The declarative language bias used in WARMR drastically simplifies the search space of all queries by using the WARMODE formalism. This formalism requires two major constraints. The most important constraint is the *key constraint*. This constraint requires the specification of a single *key* atomic formula which is obligatory in all queries. This key atomic formula also determines *what* is counted, i.e., it determines the head of the query, that is, all variables occuring in the key atom. Second, it requires a list *Atoms* of all atomic formulas that are allowed in the queries that will be generated. In the most general case, this list consists of the relation names in the database schema **R**. If one also wants to allow certain constants within the atomic formulas, then these atomic formulas must be specified for every such constant. In the most general case, the complete database instance must also be added to the *Atoms* list. The WARMODE formalism also allows other constraints, but since these are not obligatory, we will not discuss them any further.

Example 2. Consider

$$Atoms := \{\, likes(\_, \text{`Duvel'}),$$
$$likes(\_, \text{`Trappist'}),$$
$$serves(\_, \text{`Duvel'}),$$
$$serves(\_, \text{`Trappist'})\},$$

where _ stands for an arbitrary variable, and

$$key := visits(\_, \_).$$

Then,

$$\mathcal{L} = \{Q(x_1, x_2) :\!- visits(x_1, x_2), likes(x_3, \text{`Duvel'}).$$
$$Q(x_1, x_2) :\!- visits(x_1, x_2), likes(x_1, \text{`Duvel'}).$$
$$\dots$$
$$Q(x_1, x_2) :\!- visits(x_1, x_2), serves(x_3, \text{`Duvel'}).$$
$$Q(x_1, x_2) :\!- visits(x_1, x_2), serves(x_2, \text{`Duvel'}).$$
$$\dots$$
$$Q(x_1, x_2) :\!- visits(x_1, x_2), likes(x_1, \text{`Duvel'}), serves(x_2, \text{`Duvel'}).$$
$$Q(x_1, x_2) :\!- visits(x_1, x_2), likes(x_1, \text{`Duvel'}), serves(x_2, \text{`Trappist'}).$$
$$\dots\}.$$

As can be seen, these constraints already dismiss a lot of interesting patterns. However, it is still possible to discover all frequent conjunctive queries, but then, we need to run the algorithm for every possible key atomic formula with the least restrictive declarative language bias. Of course, using this strategy, a lot of possible optimizations are left out, as will be shown in the next section.

The specialization relation used in WARMR is defined $Q_1 \preceq Q_2$ if $Q_1 \subseteq Q_2$. The selection predicate q is the minimal support threshold, which is indeed monotone with respect to $\preceq$, i.e., for every instance $\mathbf{I}$ and conjunctive queries Q_1 and Q_2, we have the following: if Q_1 is frequent and $Q_1 \subseteq Q_2$, then Q_2 is frequent.

Candidate generation. In essence, the WARMR algorithm generates all conjunctive queries contained in the query $Q(\bar{x})$:- $R(\bar{x})$, where $R(\bar{x})$ is the key atomic formula. Denote this query by the *key conjunctive query*. Hence, the key conjunctive query is the (single) most general pattern in C_1. Assume we are given all frequent patterns up to a certain level i, $\bigcup_{j \leq i} \mathcal{F}_j$. Then, WARMR generates a superset of all candidate patterns, by adding a single atomic formula, from *Atoms*, to every query in $\mathcal{F}_i$, as allowed by the WARMODE declarations. From this set, every candidate pattern needs to be identified by checking whether all of its generalizations are frequent. However, this is no longer possible, since some of these generalizations might not be in the language of admissible patterns. Therefore, only those generalizations that satisfy the declarative language bias need to be known frequent. In order to do this, for each generated query Q, WARMR scans all infrequent conjunctive queries for one that is more general than Q. However, this does not imply that all queries that are more general than Q are known to be frequent! Indeed, consider the following example which is based on the declarative language bias from the previous example.

Example 3.

$$Q_1(x_1, x_2) \text{ :- } visits(x_1, x_2), likes(x_1, \text{'Duvel'}).$$
$$Q_2(x_1, x_2) \text{ :- } visits(x_1, x_2), likes(x_3, \text{'Duvel'}).$$

Both queries are single extensions of the key conjunctive query, and hence, they are generated within the same iteration. Obviously, Q_2 is more general than Q_1, but still, both queries remain in the set of candidate queries. Moreover, it is necessary that both queries remain admissible, in order to guarantee that all frequent conjunctive queries are generated.

This example shows that the candidate generation step of WARMR does not comply with the general levelwise framework given in the previous section. Indeed, at a certain iteration, it generates patterns of different levels in the search space spanned by the containment relation.

The generation strategy also generates several queries that are equivalent with other candidate queries, or with queries already generated in previous iterations, which also need to be identified and removed from the set of candidate patterns. Again, for each candidate query, all other candidate queries and all frequent queries are scanned for an equivalent query. Unfortunately, the question whether two conjunctive queries are equivalent is an NP-complete problem. Note that isomorphic queries are definitely equivalent (but not vice versa in general), and also the problem of efficiently generating finite structures up to isomorphism, or testing isomorphism of two given finite structures efficiently, is still an open problem [7].

Candidate evaluation. Since WARMR is an inductive logic programming algorithm written within a logic programming environment, the evaluation of all candidate queries is performed inefficiently. Still, WARMR uses several optimizations to increase the performance of this evaluation step, but these optimizations can hardly be compared to the optimized query processing capabilities of relational database systems.

4.2 Phase 2

The procedure to generate all association rules in WARMR, simply consists of finding all couples (Q_1, Q_2) in the list of frequent queries, such that Q_2 is contained in Q_1. We were unable to find how this procedure exactly works, that is, how is each query Q_2 found, given query Q_1. Anyhow, in general, this phase is less of an efficiency issue, since the supports of all queries that need to be considered are already known.

5 Getting WARMeR

Inspired by the framework of WARMR, we present in this section a more general framework and investigate the efficiency challenges described in Section 3. More specifically, we want to discover association rules over all conjunctive queries instead of only those queries contained in a given key conjunctive query since it might not always be clear what exactly needs to be counted. For example, in the beer drinkers database, the examples given in section 2 show that different heads could lead to several interesting association rules about the drinkers, the bars or the beers separately. We also want to exploit the containment relationship of conjunctive queries as much as possible, and avoid situations such as described in example 3. Indeed, the WARMR algorithm does not fully exploit the different levels induced by the containment relationship, since it generates several candidate patterns of different levels within the same iteration.

5.1 Phase 1

The goal of this first phase is to find all frequent conjunctive queries. Hence, $\mathcal{L}$ is the family of all conjunctive queries.

Since only the number of different tuples in the answer of a query is important and not the content of the answer itself, we will extend the notion of query containment, such that it can be better exploited in the levelwise algorithm.

Definition 4. *A conjunctive query Q_1 is diagonally contained in Q_2 if Q_1 is contained in a projection of Q_2. We write $Q_1 \subseteq^\Delta Q_2$.*

Example 4.

$$Q_1(x) :\!- likes(x, y), visits(x, z), serves(z, y)$$
$$Q_2(x, z) :\!- likes(x, y), visits(x, z), serves(z, y)$$

The answer of Q_1 consists of all drinkers that visit at least one bar that serve at least one beer they like. The answer of Q_2 consists of all visits of a drinker to a bar if that bar serves at least one beer the drinker likes. Obviously, a drinker could visit multiple bars that serve a beer they like, and hence all these bars will be in the answer of Q_2 together with that drinker, while Q_1 only gives the name of that drinker, and hence, the number of tuples in the answer of Q_1 will always be smaller or equal than the number of tuples in the answer of Q_2.

We now define $Q_1 \preceq Q_2$ if $Q_1 \subseteq^\Delta Q_2$. The selection predicate q is the minimal support threshold, which is indeed monotone with respect to $\preceq$, i.e., for every instance $\mathbf{I}$ and conjunctive queries Q_1 and Q_2, we have the following: if Q_1 is frequent and $Q_1 \subseteq^\Delta Q_2$, then Q_2 is frequent. Notice that the notion of diagonal containment now allows the incorporation of conjunctive queries with different heads within the search space spanned by this specialization relation.

Two issues remain to be solved: how are the candidate queries efficiently generated without generating two equivalent queries? and how is the frequency of each candidate query efficiently computed?

Candidate generation. As a first optimization towards the generation of all conjunctive queries, we will already prune several queries in advance.

1. The head of a query must contain at least one variable, since the support of a query with an empty head can be at most 1. Hence, we already know its support after we evaluate a query with the same body but a nonempty head.
2. We allow only a single permutation of the head, since the supports of queries with an equal body but different permutations of the head are equal.

Generating candidate conjunctive queries using the levelwise algorithm requires an initial set of all most general queries with respect to $\subseteq^\Delta$. However, such queries do not exist. Indeed, for every conjunctive query Q, we can construct another conjunctive query Q', such that $Q \subseteq^\Delta Q'$ by simply adding a new atomic formula with new variables into the body of Q, and adding these variables to the head. A rather drastic but still reasonable solution to this problem is to apriori limit the search space to conjunctive queries with at most a fixed number of atomic formulas in the body. Then, within this space, we can look at the set of most general queries, and this set now is well-defined.

At every iteration in the levelwise algorithm we need to generate all candidate conjunctive queries up to equivalence, such that all of their generalizations are known to be frequent. Since an algorithm to generate exactly this set is not known, we will generate a small superset of all candidates and afterwards remove each query of which a generalization is not known to be frequent (or known to be infrequent).

Nevertheless, any candidate conjunctive query is always more specific than at least one query in $\mathcal{F}_i$. Hence, we can generate a superset of all possible candidate queries using the following four operations on each query in $\mathcal{F}_i$.

Extension: We add a new atomic formula with new variables to the body.

Join: We replace all occurrences of a variable with another variable already occurring in the query.

Selection: We replace all occurrences of a variable x with some constant.

Projection: We remove a variable from the head if this does not result in an empty head.

Example 5. This example shows a single application of each operation on the query

$$Q(x,y) :- likes(x,y), visits(x,z), serves(z,u).$$

Extension:

$$Q(x,y) :- likes(x,y), visits(x,z), serves(z,u), likes(v,w).$$

Join:

$$Q(x,y) :- likes(x,y), visits(x,z), serves(z,y).$$

Selection:

$$Q(x,y) :- likes(x,y), visits(x,z), serves(z, \text{'Duvel'}).$$

Projection:

$$Q(x) :- likes(x,y), visits(x,z), serves(z,u).$$

The following proposition implies that if we apply a sequence of these four operations on the current set of frequent conjunctive queries, we indeed get at least all candidate queries.

Proposition 2. $Q_1 \subseteq^\Delta Q_2$ if and only if a query equivalent to Q_1 can be obtained from Q_2 by applying some finite sequence of extension, join, selection and projection operations.

Nevertheless, using these operations, several equivalent or redundant queries can be generated. An efficient algorithm avoiding the generation of equivalent queries is still unknown. Hence, whenever we generate a candidate query, we need to test whether it is equivalent with another query we already generated. In order to keep the generated superset of all candidate conjunctive queries as small as possible, we apply an operator once on each query. If the query is redundant or equivalent with a previously generated query, we repeatedly apply any of the operators until a query is found that is not equivalent with a previously generated query. As already mentioned in the previous section, testing equivalence cannot be done efficiently.

After generating this superset of all candidate conjunctive queries, we need to check for each of them whether all more general conjunctive queries are known to be frequent. This can be done by performing the inverses of the four operations extension, join, selection and projection, as described above. Even if we now assume that in the set of all frequent conjunctive queries there exist no two equivalent queries, we still need to find the query equivalent to the one generated using the inverse operations. Hence, the challenge of testing equivalence of two conjunctive queries reappears.

Candidate evaluation. After generating all candidate conjunctive queries, we need to test which of them are frequent. This can be done by simply evaluating every candidate query on the database, one at a time, by translating each query to SQL. Although conjunctive query evaluation in relational databases is well understood and several efficient algorithms have been developed (i.e., join query optimisation and processing) [8], this remains a costly operation. Within database research, a lot of research has been done on multi-query optimization [15]. Here, one tries to efficiently evaluate multiple queries at once. Unfortunately, these techniques are not yet implemented in most common database systems.

As a first optimization towards query evaluation, we can already derive the support of a significant part of all candidate conjunctive queries. Therefore, we only consider those candidate queries that satisfy the following restrictions.

1. We only consider queries that have no constants in the head, because the support of such queries is equal to the support of those queries in which the constant is not in the head.
2. We only consider queries that contain no duplicate variables in the head, since the support of such a query is equal to the support of the query without duplicates in the head.

As another optimization, given a query involving constants, we will not treat every variation of that query that uses different constants as a separate query, but rather we can evaluate all those variations in a single global query. For example, suppose the query

$$Q(x_1) \; :- \; R(x_1, x_2)$$

is frequent. From this query, a lot of candidate queries are generated using the selection operation on x_2. Assume the active domain of x_2 is $1, 2, \ldots, n$, then the set of candidate queries contains at least

$$\{Q(x_1) \; :- \; R(x_1, 1), Q(x_1) \; :- \; R(x_1, 2), \ldots, Q(x_1) \; :- \; R(x_1, n)\},$$

resulting in a possibly huge amount of queries that need to be evaluated. However, the support of all these queries can be computed by evaluating only the single SQL query

> **select** x_2, **count**(∗)
> **from** R
> **group by** x_2
> **having count**(∗) $\geq$ *minsup*

of which the answer consists of every possible constant c for x_2 together with the support of the corresponding query $Q(x_1) \; :- \; R(x_1, c)$. From now on, we will therefore use only a symbolic constant to denote all possible selections of a given variable. For example, $Q(x_1) \; :- \; R(x_1, c_1)$ denotes the set of all possible selections for x_2 in the previous example. A query with such a symbolic constant is then considered frequent if it is frequent for at least one constant.

As can be seen, several optimizations can be used to improve the performance of the evaluation step in our algorithm. Also, we might be able to use some of the techniques that have been developed for frequent itemset mining, such as closed frequent itemsets [14], free sets [2] and non derivable itemsets [3]. These techniques could then be used to minimize the number of candidate queries that need to be executed on the database, but instead we might be able to compute their supports based on the support of previously evaluated queries. Another interesting optimization could be to avoid using SQL queries completely, but instead use a more intelligent counting mechanism that needs to scan the database or the materialized tables only once, and count the supports of all queries at the same time.

5.2   Phase 2

The goal of the second phase is to find for every frequent conjunctive query Q, all confident association rules $Q' \Rightarrow Q$. Hence, we need to run the general levelwise algorithm separately for every frequent query. That is, for any given Q, $\mathcal{L}$ consists of all conjunctive queries Q', such that $Q \subseteq Q'$. Assume we are given two association rules $AR_1 : Q_1 \Rightarrow Q_2$ and $AR_2 : Q_3 \Rightarrow Q_4$, we define $AR_1 \preceq AR_2$ if $Q_3 \subseteq Q_1$ and $Q_2 \subseteq Q_4$.

The selection predicate q is the minimal confidence threshold which is again monotone with respect to $\preceq$, i.e., for every instance $\mathbf{I}$ and association rules $AR_1 : Q_1 \Rightarrow Q_2$ and $AR_2 : Q_3 \Rightarrow Q_4$, we have the following: if AR_1 is frequent and confident and $AR_1 \preceq AR_2$, then AR_2 is frequent and confident.

Here, only a single issue remains to be solved: how are the candidate queries efficiently generated without generating two equivalent queries?

We have to generate, for every frequent conjunctive query Q, all conjunctive queries Q', such that $Q \subseteq Q'$ and minimize the generation of equivalent queries. In order to do this, we can use three of the four inverse operations described for the previous phase, i.e., the inverse extension, inverse join and inverse selection operations. We do not need to use the inverse projection operation since we do not want those queries that are diagonally contained in Q, but only those queries that are regularly contained in Q as defined in Section 2. Still, several queries will be generated which are equivalent with previously generated queries, and hence this should again be tested.

6   Sample Run

Suppose we are given an instance of the beer drinkers database used throughout this paper, as shown in Figure 1.

We now show a small part of an example run of the algorithm presented in the previous section. In the first phase, all frequent conjunctive queries need to be found, starting from the most general conjunctive queries. Let the maximum number of atoms in de body of the query be limited to 2, and let the minimal support threshold be 2, i.e., at least 2 tuples are needed in the output of a query

Likes	
Drinker	*Beer*
Allen	Duvel
Allen	Trappist
Carol	Duvel
Bill	Duvel
Bill	Trappist
Bill	Jupiler

Visits	
Drinker	*Bar*
Allen	Cheers
Allen	California
Carol	Cheers
Carol	California
Carol	Old Dutch
Bill	Cheers

Serves	
Bar	*Beer*
Cheers	Duvel
Cheers	Trappist
Cheers	Jupiler
California	Duvel
California	Jupiler
Old Dutch	Trappist

Fig. 1. Instance of the beer drinkers database.

to be considered frequent. Then, the initial set of candidate queries C_1, consists of the 6 queries as shown in Figure 2.

$$Q_1(x_1, x_2, x_3, x_4) :- likes(x_1, x_2), likes(x_3, x_4)$$
$$Q_2(x_1, x_2, x_3, x_4) :- likes(x_1, x_2), visits(x_3, x_4)$$
$$Q_3(x_1, x_2, x_3, x_4) :- likes(x_1, x_2), serves(x_3, x_4)$$
$$Q_4(x_1, x_2, x_3, x_4) :- visits(x_1, x_2), visits(x_3, x_4)$$
$$Q_5(x_1, x_2, x_3, x_4) :- visits(x_1, x_2), serves(x_3, x_4)$$
$$Q_6(x_1, x_2, x_3, x_4) :- serves(x_1, x_2), serves(x_3, x_4)$$

Fig. 2. Level 1.

Obviously, the support of each of these queries is 36, and hence, $F_1 = C_1$. To generate all candidate conjunctive queries for level 2, we need to apply the four specialization operations to each of these 6 queries. Obviously, the extension operation is not yet allowed, since this would result in a conjunctive queries with 3 atoms in their bodies. We can apply the Join operation on Q_1, resulting in queries Q_7 and Q_8, as shown in Figure 3. Similarly, the join operation can be applied to Q_4 and Q_6, resulting in queries Q_9, Q_{10} and Q_{11}, Q_{12} respectively. However, the Join operation is not allowed on Q_2, Q_3 and Q_5, since for each of them, there always exists a query in which it is contained and which is not yet known to be frequent. For example, if we join x_1 and x_3 in query Q_2, resulting in $Q(x_1, x_2, x_3) :- likes(x_1, x_2), visits(x_1, x_3)$, then this query is contained in $Q'(x_1, x_2, x_4) :- likes(x_1, x_2), visits(x_3, x_4)$, of which the frequency is not yet known. Similar situations occur for the other possible joins on Q_2, Q_3 and Q_5. The selection operation can also not be applied to any of the queries, since for each variable we would select, there always exists a more general query in which that variable is projected, but not selected, and hence, the frequency of such queries is yet unknown. We can apply the projection operator on any variable of queries Q_1 through Q_6, resulting in queries Q_{13} to Q_{37}.

In stead of showing the next levels for all possible queries, we will show only single path, starting from query Q_7. On this query, we can now also apply the projection operation on x_3. This results in a redundant atom which can be removed, resulting in the query $Q_7'(x_1, x_2) :- likes(x_1, x_2)$. Again, for the next level, we

$$Q_7(x_1, x_2, x_3) \;:\!\!-\; likes(x_1, x_2), likes(x_1, x_3)$$
$$Q_8(x_1, x_2, x_3) \;:\!\!-\; likes(x_1, x_2), likes(x_2, x_3)$$
$$Q_9(x_1, x_2, x_3) \;:\!\!-\; visits(x_1, x_2), visits(x_1, x_3)$$
$$Q_{10}(x_1, x_2, x_3) \;:\!\!-\; visits(x_1, x_2), visits(x_2, x_3)$$
$$Q_{11}(x_1, x_2, x_3) \;:\!\!-\; serves(x_1, x_2), serves(x_1, x_3)$$
$$Q_{12}(x_1, x_2, x_3) \;:\!\!-\; serves(x_1, x_2), serves(x_2, x_3)$$
$$Q_{13}(x_2, x_3, x_4) \;:\!\!-\; likes(x_1, x_2), likes(x_3, x_4)$$
$$\vdots$$
$$Q_{37}(x_1, x_2, x_3) \;:\!\!-\; serves(x_1, x_2), serves(x_3, x_4)$$

Fig. 3. Level 2.

can use the projection operation on x_2, now resulting in $Q_7''(x_1) :\!\!- likes(x_1, x_2)$. Then, for the following level, we can use the selection operation on x_2, resulting in the query $Q_7'''(x_1) :\!\!- likes(x_1, \text{'Duvel'})$. Note that if we had selected x_2, using the constant 'Trappist', then the resulting query would not have been frequent and would have been removed for further consideration. If we repeatedly apply the four specialization operations until the levelwise algorithm stops, because no more candidate conjunctive queries could be generated anymore, the second phase can start generating confident association rules from all generated frequent conjunctive queries. For example, starting from query Q_7'', we can apply the inverse selection operation, resulting in Q_7''. Since both these queries have support 3, the rule $Q_7'' \Rightarrow Q_7'''$ holds with 100% confidence, meaning that every drinker that likes a beer, also likes Duvel, according to the given database.

7   Conclusions and Future Research

In the future, we plan to study subclasses of conjunctive queries for which there exist efficient candidate generation algorithms up to equivalence. Possibly interesting classes are conjunctive queries on relational databases that consist of only binary relations. Indeed, every relational database can be decomposed into a database consisting of only binary relations. If necessary, this can be further simplified by only considering those conjunctive queries that can be represented by a tree. Note that one of the underlying challenges that always reappears is the equivalence test, which can be computed efficiently on tree structures. Other interesting subclasses are the class of acyclic conjunctive queries and queries with bounded query-width, since also for these structures, equivalence testing can be done efficiently [4].

However, by limiting the search space to one of these subclasses, Proposition 1 is no longer valid, since the intersection of two queries within such a subclass does not necessarily result in a conjunctive query which is also in that subclass.

Another important topic is the improvement of performance issues for evaluating all candidate queries. Also the problem of allowing flexible constraints to efficiently limit the search space to an interesting subset of all conjunctive queries, is an important research topic.

References

1. R. Agrawal, H. Mannila, R. Srikant, H. Toivonen, and A.I. Verkamo. Fast discovery of association rules. In U.M. Fayyad, G. Piatetsky-Shapiro, P. Smyth, and R. Uthurusamy, editors, *Advances in Knowledge Discovery and Data Mining*, pages 307–328. MIT Press, 1996.
2. J-F. Boulicaut, A. Bykowski, and C. Rigotti. Free-sets: a condensed representation of boolean data for frequency query approximation. *Data Mining and Knowledge Discovery*, 2001. To appear.
3. T. Calders and B. Goethals. Mining all non-derivable frequent itemsets. In *Proceedings of the 6th European Conference on Principles of Data Mining and Knowledge Discovery*, Lecture Notes in Computer Science. Springer-Verlag, 2002. to appear.
4. C. Chekuri and A. Rajaraman. Conjunctive query containment revisited. *Theoretical Computer Science*, 239(2):211–229, 2000.
5. L. Dehaspe and H. Toivonen. Discovery of frequent datalog patterns. *Data Mining and Knowledge Discovery*, 3(1):7–36, 1999.
6. L. Dehaspe and H. Toivonen. Discovery of relational association rules. In S. Dzeroski and N. Lavrac, editors, *Relational data mining*, pages 189–212. Springer-Verlag, 2001.
7. S. Fortin. The graph isomorphism problem. Technical Report 96-20, University of Alberta, Edmonton, Alberta, Canada, July 1996.
8. H. Garcia-Molina, J. Ullman, and J. Widom. *database system implementation*. Prentice-Hall, 2000.
9. D. Hand, H. Mannila, and P. Smyth. *Principles of Data Mining*. MIT Press, 2001.
10. A. Inokuchi and H. Motoda T. Washio. An apriori-based algorithm for mining frequent substructures from graph data. In *Proceedings of the 4th European Conference on Principles of Data Mining and Knowledge Discovery*, volume 1910 of *Lecture Notes in Computer Science*, pages 13–23. Springer-Verlag, 2000.
11. M. Kuramochi and G. Karypis. Frequent subgraph discovery. In *Proceedings of the 2001 IEEE International Conference on Data Mining*, pages 313–320. IEEE Computer Society, 2001.
12. H. Mannila and H. Toivonen. Levelwise search and borders of theories in knowledge discovery. *Data Mining and Knowledge Discovery*, 1(3):241–258, November 1997.
13. S.H. Nienhuys-Cheng and R. de Wolf. *Foundations of Inductive Logic Programming*, volume 1228 of *Lecture Notes in Artificial Intelligence*. Springer-Verlag, 1997.
14. N. Pasquier, Y. Bastide, R. Taouil, and L. Lakhal. Discovering frequent closed itemsets for association rules. In *Proceedings of the 7th International Conference on Database Theory*, volume 1540 of *Lecture Notes in Computer Science*, pages 398–416. Springer-Verlag, 1999.
15. P. Roy, S. Seshadri, S. Sudarshan, and S. Bhobe. Efficient and extensible algorithms for multi query optimization. In *Proceedings of the 2000 ACM SIGMOD International Conference on Management of Data*, volume 29:2 of *SIGMOD Record*, pages 249–260. ACM Press, 2000.
16. J.D. Ullman. *Principles of database and knowledge-base systems, volume 2*, volume 14 of *Principles of Computer Science*. Computer Science Press, 1989.
17. M. Zaki. Efficiently mining frequent trees in a forest. In *Proceedings of the Eight ACM SIGKDD International Conference on Knowledge Discovery and Data Mining*. ACM Press, 2002. to appear.

Mining Text Data: Special Features and Patterns

M. Delgado, M.J. Martín-Bautista, D. Sánchez*, and M.A. Vila

Department of Computer Science and Artificial Intelligence
University of Granada, Spain

Abstract. Text mining is an increasingly important research field because of the necessity of obtaining knowledge from the enormous number of text documents available, especially on the Web. Text mining and data mining, both included in the field of information mining, are similar in some sense, and thus it may seem that data mining techniques may be adapted in a straightforward way to mine text. However, data mining deals with structured data, whereas text presents special characteristics and is basically unstructured. In this context, the aims of this paper are three:

- To study particular features of text.
- To identify the patterns we may look for in text.
- To discuss the tools we may use for that purpose.

In relation with the third point we overview existing proposals, as well as some new tools we are developing by adapting data mining tools previously developed by our research group.

1 Introduction

Nowadays, there is a large amount of information available in the form of text in diverse environments. The analysis of such data can provide many benefits in several areas, and that has motivated an increasing interest in developing technologies for that purpose. In this context, the modern paradigm called Text Mining copes with the problem of finding interesting patterns in textual data.

Text mining is usually considered as a subdomain of data mining, a field concerned with finding interesting patterns in databases, but in the setting of text. In the same way data mining is part of the global process of Knowledge Discovery in Databases, text mining could be seen as part of the Knowledge Discovery in Texts (KDT) paradigm. However, as is also the case of data mining, there are several different interpretations of these terms [17,21,23]. In many occasions, the "mining" process is not distinguished from the global KDT one.

What gives text mining a clear distinction from data mining is that most of the times, the latter deals with structured data, whereas text presents special characteristics and its explicit appearance seems to be basically unstructured. However, textual data is not inherently unstructured. On the contrary, text

---

* Corresponding author. daniel@decsai.ugr.es, Phone: +34 958 246397, Fax: +34 958 243317.

D.J. Hand et al. (Eds.): Pattern Detection and Discovery, LNAI 2447, pp. 140–153, 2002.

is characterized by a very complex implicit structure that has defeated many representation attempts. The problem is then to design automatic tools able to make explicit the implicit structure of textual data.

More than being an inconvenient, this characteristic of textual data opens a wide range of possibilities and raises many interesting issues that exceed those of data mining on structural data. In particular, the definition of specific patterns for text and the problem of adapting classical patterns have motivated a big research effort in recent years.

In this context, our goal is to study some particular features and patterns in textual data, as well as both existing and potential tools in text mining. Of course, this overview is not intended to be exhaustive. Despite the youth of this field, the literature is very broad and keeps growing in an impressive way.

2   Special Features of Textual Data

2.1   Implicit Structure

Most of the traditional work on data analysis, including statistics, machine learning and data mining, assume data is structured. Typically, this structure consists of a collection of cases, each case described by a certain set of variables.

On the contrary, data appearing in the form of text is usually seen in the literature as unstructured data. Because of that, a first step in text mining is to transform text into structured data. This is called *text refining* in [43], and plays a central role in text mining processes.

It is important to note, however, that textual data appears in an implicitly structured, rather than unstructured, form [36]. Each text document is an ordered collection of words and punctuation signs, with attached meanings, whose situation in the text is ruled by complex syntactic and semantic restrictions. In addition, semi-structured text such as XML documents offer some additional features and relations that can be exploited.

The complexity of the implicit structure of text can be appreciated by looking at some of the classical models that attempts to represent it (either total or partially), such as conceptual graphs [38] and scripts [39]. This is also acknowledged in [37] in relation to information retrieval tasks.

A related and important issue here is the necessity of managing background knowledge in complex problems, usually about an specific application domain. This background knowledge involves different kinds of relations among terms and concepts, like those represented by thesauri and ontologies. Also, frame descriptions of the expected structure of documents in a given application domain fall into this category.

2.2   Text Refining

Text refining can then be defined as the process of making explicit the implicit structure of text, by mapping it to a (much simpler) structured representation, called *intermediate form* in [43].

The implicit structure of textual data is so rich that it can be made explicit in many different ways. This is acknowledged by the large amount of different intermediate forms proposed in the literature. Among the text features considered when building an intermediate form from text we can cite words, terms, keywords (all of them well-known from information retrieval), events [22] and other information extraction structures [32], XML tags [46], episodes [3], etc.

Extracting such features from text benefits from the existing techniques of Natural Language Processing (NLP), whose development come motivated by older research areas such as Information Retrieval and Computational Linguistics, and more recent ones like Information Extraction, among others [21].

The question is, what intermediate form may we use? Quoting Feldman and Dagan [17]

> The [explicit] structure should reflect the way in which the user conceptualize the domain that is described by the data.

In practice, this means intermediate forms should be related to the kind of patterns we are interested in. However, this is not the only criterion. The structuring process has to be made automatically and sometimes, to obtain complex structures requires an unreasonable computational cost [17]. Even more, it could happen that no robust algorithm exists to extract certain kind of structures.

In the next sections we shall overview some intermediate forms and several kinds of patterns we can obtain from them.

3 Association Rules

One of the first data mining techniques employed in mining text collections is that of finding association rules (AR's) [1]. Discovering AR's in text collections is very interesting since many applications related to text processing involve associations and co-occurrences between text features.

As it is well known, AR's are "implications" that relate the presence of itemsets in transactions. Then, in a first view, all we need to start searching for AR's is a set T of transactions (T-set for short), each transaction being a subset of a set of items I. However, there are many ways to obtain a T-set from a collection of texts, and consequently different kinds of associations can be derived[1]. In addition, there are several patterns related to AR's with particular features. Most of our work in this paper focuses on these topics.

3.1 Association Rules in Textual Data

Given a set of items I and a T-set T on I, an association rule is an expression of the form $A \Rightarrow C$ with $A, C \subseteq I$, $A, C \neq \emptyset$ and $A \cap C = \emptyset$, where it is usual to consider $|C| = 1$. The meaning of such rule is in general "C is in every transaction where A is".

---

[1] This happens even in the case of relational databases, see [15,9] for an interpretation of approximate dependencies as association rules.

Some papers have been devoted to association extraction in collections of texts. The main difference between them is the intermediate form they employ.

- A first approach looks at text documents as bags of words. In that case, the simplest way to obtain a T-set from a collection of texts is to consider only one occurrence of each word in each document. Then, I is the set of words appearing in all the collection, and T is the set of documents. However, I becomes too large, and many of the words in I could be uninformative or uninteresting, so this approach does not provide suitable results [36].
- To avoid this, automated indexing techniques from NLP field are employed to select only those key-words that best describe the documents. Association extraction from indexed data is described in [18,20,36] among others.
  In this approach, $I = \{k_1, \ldots, k_m\}$ is a set of keywords employed to label a set of documents. Then, each document becomes a transaction, i.e., $T = \{d_1, \ldots, d_n\}$, and rules take the form $\{k_{i_1}, \ldots, k_{i_p}\} \Rightarrow k_{i_{p+1}}$, where $p \leq m$. The meaning of such rules is "every document labelled with $\{k_{i_1}, \ldots, k_{i_p}\}$ is also labelled with $k_{i_{p+1}}$".
  A slight variation of this scheme is to consider I to be compound of terms instead of keywords. The main difference is that a keyword is representative of a concept, while a term (word or phrase) is probably not [19,26].
- Generalized Association Rules [42] make use of a specific kind of background knowledge, namely an item taxonomy, to discover associations at different granularity levels. This kind of background knowledge has been employed for example in the FACT system [20] and in [19] to discover associations between terms, given a term taxonomy.
- In [5], items are multi-term text phrases, and transactions are documents. The goal is to discover co-occurring phrases in documents. A more elaborated idea is described in [3], where phrases are generalized to *episodes* [28]. Roughly speaking, an episode is a set of text feature vectors and an associated order relation. The rules that relate episodes are called here *episode rules*.

Further generalizations of association rules for text mining purposes are described in [36].

3.2 Assessing Association Rules

One of the generalizations we find more interesting among those in [36] is to provide different quality measures as an alternative to the usual support/confidence framework.

For example, Kodratoff [23] distinguishes different semantics of the implication in an association rule. These are *descriptive* rules, that serve to describe properties of an object, and *causal* rules, where the implication carries a causal meaning. A certain number of measures, associated to more specific semantics, are provided. This is an interesting research issue for text mining because there are many different kinds of features and relations in text data, each one with its own particular semantics.

The support/confidence framework. The ordinary measures proposed in [1] are *confidence* and *support*, both based on the support of an itemset (a set of items). The support of an itemset is the percentage of transactions where the itemset appears. Support of a rule is the percentage of tuples where the rule holds. Confidence is defined as

$$Conf(A \Rightarrow C) = \frac{supp(A \cup C)}{supp(A)} = \frac{Supp(A \Rightarrow C)}{supp(A)} \, . \tag{1}$$

However, several authors have pointed out some drawbacks of this framework (see [11,41,8]). The main ones are:

- Confidence does not satisfy the three requirements stated by Piatetsky-Shapiro [35] for any suitable accuracy measure. In particular, it is not able to detect either statistical independence or negative dependence between antecedent and consequent.
- When $supp(C)$ is very high and $Conf(A \Rightarrow C) > supp(C)$, it is very likely that a big amount of misleading rules with very high confidence are reported.

In the following we describe our approach to assess association rules. Further details can be found in [8].

Certainty factors. We have recently proposed Shortliffe and Buchanan's certainty factors [40] as an alternative to confidence [8]. The certainty factor of a rule is defined as

$$CF(A \Rightarrow C) = \frac{(Conf(A \Rightarrow C)) - supp(C)}{1 - supp(C)} \tag{2}$$

if $Conf(A \Rightarrow C) > supp(C)$, and

$$CF(A \Rightarrow C) = \frac{(Conf(A \Rightarrow C)) - supp(C)}{supp(C)} \tag{3}$$

if $Conf(A \Rightarrow C) < supp(C)$, and 0 otherwise.

Certainty factors take values in $[-1, 1]$, indicating the extent to which our belief that the consequent is true varies when the antecedent is also true. It ranges from 1, meaning maximum increment (i.e., when A is true then B is true) to -1, meaning maximum decrement.

They satisfy the requirements stated by Piatetsky-Shapiro [35] for any suitable accuracy measure. In particular, they are able to check statistical independence between A and B, and in that case $CF(A \Rightarrow B) = 0$. In addition they are able to detect negative dependence and, in that case, $CF(A \Rightarrow B) < 0$. Further properties can be found in [8].

Very strong rules. We introduced the concept of very strong rule [8] in the following way: the rule $A \Rightarrow C$ is very strong iff both $A \Rightarrow C$ and $\neg C \Rightarrow \neg A$ are strong rules.

The rationale behind this definition is that $A \Rightarrow C$ and $\neg C \Rightarrow \neg A$ are logically equivalent, so we should look for strong evidence of both rules in data to believe that they are interesting.

This definition can help us to solve the high-support itemsets drawback since when $supp(C)$ (or $supp(A)$) is very high, $Supp(\neg C \Rightarrow \neg A)$ is very low, and hence the rule $\neg C \Rightarrow \neg A$ won't be strong and $A \Rightarrow C$ won't be very strong.

Several properties of this definition can be found in [8]. In particular, the following holds for positive certainty factors:

- Let $supp(A) + supp(C) \geq 1$. Then $A \Rightarrow C$ is very strong iff $A \Rightarrow C$ is strong.
- Let $supp(A) + supp(C) \leq 1$. Then $A \Rightarrow C$ is very strong iff $\neg C \Rightarrow \neg A$ is strong.

In practice, when searching for very strong rules we are searching for rules with support over a threshold *minsupp* and under a threshold *maxsupp* that is at most 1-*minsupp* [8]. This way we avoid the problems introduced by itemsets with very high support.

Let us remark that this new framework can be easily employed in most of the existing algorithms to discover association rules, without increasing neither time nor space complexity. This benefits from the following properties:

- If $CF(A \Rightarrow C) > 0$ then $CF(A \Rightarrow C) = CF(\neg C \Rightarrow \neg A) > 0$. Hence, we must check only one CF condition.
- $Supp(\neg C \Rightarrow \neg A) = 1 - supp(C) - supp(A) + Supp(A \Rightarrow C)$. Since $supp(C)$, $supp(A)$ and $Supp(A \Rightarrow C)$ are available after the first step of the algorithms, checking the support condition $Supp(\neg C \Rightarrow \neg A) > minsupp$ is inmediate.

Related issues. We think there are some interesting issues related to the assessment of association rules on transactions obtained from text, that we'll investigate in the future.

- A first relation concerns document indexing and the selection of relevant terms. One of the well-known measures involved in this process, inverse document frequency (IDF) [37], is related to the support of terms (items) in the set of documents (transactions). Indeed, the problem addressed by IDF is that of detecting that the support of a term in a given collection is too high, and this is closely related to our discussion about itemsets with very high support, and the definition of very strong rule.
- In relation to the same topic we are also concerned with using term weights, similar to those provided by document indexing, when searching for rules relating terms. A general framework that fits this problem is that of weighted items and weighted rules [45]. In section 4 we shall describe another kind of patterns, fuzzy association rules, able to take into account a fuzzy weighting of the terms.

We think that using the appropriate framework we could find an alternative to document indexing in order to avoid finding uninteresting associations when we consider the entire set of terms. Some related works in term filtering [19] and feature selection [31,30] are available.

4   Fuzzy Association Rules

In many cases, text features that are identified with items are in a transaction to a certain degree. For example, terms in a document are provided by indexing techniques with a relevance degree.

The computation and semantics of that value differ from one technique to another, but what is accepted in many text-related areas such as information retrieval, is that taking into account such information improves the performance of a system. Hence, it seems natural to consider those degrees also in text mining applications. For example, one method that finds rules and is based on the vector model and term frequencies is described in [33].

One of the existing methods to compute the relevance degree of a term is based on fuzzy weights. To make it possible to perform a text mining task taking into account that information, we are interested in using fuzzy association rules.

In this section we describe the definition of fuzzy association rule (FAR) we introduced in [12], and one of its possible applications in text mining.

4.1   Fuzzy Transactions and *FT*-Sets

The starting point in our definition of FAR is the concept of fuzzy transaction. Given a set of items I, we call fuzzy transaction any nonempty fuzzy subset $\tilde{\tau} \subseteq I$. For every $i \in I$, $\tilde{\tau}(i)$ is the membership degree of i in a fuzzy transaction $\tilde{\tau}$.

We call a (crisp) set of fuzzy transactions a *FT*-set. *FT*-sets can be represented by means of a table, where columns and rows are labelled with identifiers of items and transactions respectively. The cell for item i_k and transaction $\tilde{\tau}_j$ contains the value $\tilde{\tau}_j(i_k)$.

The following example is from [12]. Let $I = \{i_1, i_2, i_3, i_4\}$ be a set of items. Table 1 describes three transactions defined on items of I.

Table 1. Three fuzzy transactions

	i_1	i_2	i_3	i_4
$\tilde{\tau}_1$	0	0.6	1	1
$\tilde{\tau}_2$	0	1	0	1
$\tilde{\tau}_3$	1	0.4	0.75	0.1

Here, $\tilde{\tau}_1 = 0.6/i_2 + 1/i_3 + 1/i_4$, $\tilde{\tau}_2 = 1/i_2 + 1/i_4$ and $\tilde{\tau}_3 = 1/i_1 + 0.4/i_2 + 0.75/i_3 + 0.1/i_4$.

We can focus on the role of a given item in a certain *FT*-set by means of the concept of *representation of an item*. The representation of an item i_k in a *FT*-set T is a fuzzy subset of T defined as follows:

$$\widetilde{\Gamma}_{i_k} = \sum_{\tilde{\tau}_j \in T} \tilde{\tau}_j(i_k)/\tilde{\tau}_j$$

In the previous example we have, for example

$$\widetilde{\Gamma}_{\{i_1\}} = 1/\tilde{\tau}_3$$
$$\widetilde{\Gamma}_{\{i_2\}} = 0.6/\tilde{\tau}_1 + 1/\tilde{\tau}_2 + 0.4/\tilde{\tau}_3$$

$$\vdots$$

We can extend the representation of an item to itemsets in the following way:

Definition 1. *The representation of an itemset $I_0 \subseteq I$ in a FT-set T is*

$$\widetilde{\Gamma}_{I_0} = \bigcap_{i \in I_0} \widetilde{\Gamma}_i$$

We use the minimum for the intersection. In our example we can find, among others

$$\widetilde{\Gamma}_{\{i_2,i_4\}} = 0.6/\tilde{\tau}_1 + 1/\tilde{\tau}_2 + 0.1/\tilde{\tau}_3$$
$$\widetilde{\Gamma}_{\{i_1,i_3,i_4\}} = 0.1/\tilde{\tau}_3$$

$$\vdots$$

4.2 Association Rules in *FT*-Sets

Let I be a set of items and let T be an FT-set based on I. Then, a fuzzy association rule is a link of the form $A \Rightarrow C$ such that $A, C \subset I$ and $A \cap C = \emptyset$.

The meaning of a fuzzy association rule is the usual for association rules, i.e., C appears in every transaction where A is. The only difference is that the set of transactions where the rule holds is a *FT*-set, and hence both A and C are in every fuzzy transaction $\tilde{\tau} \in T$ to a certain degree (given by $\widetilde{\Gamma}_A$ and $\widetilde{\Gamma}_C$ specifically). This leads to the following proposition:

Proposition 1. *([12]) A fuzzy association rule $A \Rightarrow C$ holds with total accuracy in T when $\widetilde{\Gamma}_A \subseteq \widetilde{\Gamma}_C$.*

Let us remark that ordinary transactions and AR's are special cases of fuzzy transactions and FAR's.

4.3 Assessing Fuzzy Association Rules

We shall use a semantic approach based on the evaluation of quantified sentences [49]. A quantified sentence is an expression of the form "Q of F are G", where F and G are two fuzzy subsets on a finite set X, and Q is a relative fuzzy quantifier. Relative quantifiers are linguistic labels for fuzzy percentages that can be represented by means of fuzzy sets on $[0, 1]$, such as "most", "almost all" or "many".

An example is "many young people are tall", where $Q = many$, and F and G are possibility distributions induced in the set $X = people$ by the imprecise terms "young" and "tall" respectively. An special case of quantified sentence appears when $F = X$, as in "most of the terms in the profile are relevant". The evaluation of a quantified sentence yields a $[0, 1]$-value, that assesses the accomplishment degree of the sentence.

In [12], a family of operators to obtain the support and confidence of fuzzy association rules are defined, taking Q as a parameter. Here we choose the quantifier Q_M, defined by the membership function $Q_M(x) = x$. Using this quantifier, we guarantee that in the crisp case, the following definitions yield the correct support and confidence of AR's.

Definition 2. *Let $I_0 \subseteq I$. The support of I_0 in T is the evaluation of the quantified sentence*

$$Q_M \ of \ T \ are \ \widetilde{\Gamma}_{I_0}$$

Let $A, C \subseteq I$, $A, C \neq \emptyset$, $A \cap C = \emptyset$.

Definition 3. *The support of the fuzzy association rule $A \Rightarrow C$ in the set of fuzzy transactions T is supp($A \cup C$), i.e., the evaluation of the quantified sentence*

$$Q_M \ of \ T \ are \ \widetilde{\Gamma}_{A \cup C} = \ Q_M \ of \ T \ are \ \left(\widetilde{\Gamma}_A \cap \widetilde{\Gamma}_C \right)$$

Definition 4. *The confidence of the fuzzy association rule $A \Rightarrow C$ in the set of fuzzy transactions T is the evaluation of the quantified sentence*

$$Q_M \ of \ \widetilde{\Gamma}_A \ are \ \widetilde{\Gamma}_C$$

Certainty factors are obtained from confidence of the rule and support of C in the way we detailed in previous sections.7

We evaluate the sentences by means of method GD [16], that has been shown to satisfy good properties with better performance than others. The evaluation of "Q of F are G" by means of GD is defined as

$$GD_Q(G/F) = \sum_{\alpha_i \in \Delta(G/F)} (\alpha_i - \alpha_{i+1}) Q \left(\frac{|(G \cap F)_{\alpha_i}|}{|F_{\alpha_i}|} \right) \qquad (4)$$

where $\Delta(G/F) = \Lambda(G \cap F) \cup \Lambda(F)$, $\Lambda(F)$ being the level set of F, and $\Delta(G/F) = \{\alpha_1, \ldots, \alpha_p\}$ with $\alpha_i < \alpha_{i+1}$ for every $i \in \{1, \ldots, p\}$. The set F is assumed to be normalized. If not, F is normalized and the normalization factor is applied to $G \cap F$.

In [12] we show how to adapt existing algorithms for mining AR's in order to mine for FAR's in fuzzy transactions. A valuable feature of the proposed methodology is that the complexity of the algorithms does not increase.

4.4   Applications

Our study of FAR's for text mining has been motivated by a couple of applications. The first one is to provide a tool to particularize or generalize queries in collections of documents [13] under the fuzzy retrieval model [10,24]. Work on using term associations for query expansion with crisp terms has been conducted in [34,44,47].

In our view, transactions are documents, and items are terms appearing in them. Items have associated a fuzzy relevance degree, so that each document can be seen as a fuzzy subset of items, i.e., a fuzzy transaction. The entire collection is then an FT-set.

The idea in this application is that a fuzzy rule $A \Rightarrow C$ with enough accuracy can serve to particularize a query involving term C, by replacing C with A. For the particularization to be efficient it is necessary that $C \Rightarrow A$ does not have a high accuracy, because otherwise A and C are interchangeable without modifying the set of documents retrieved.

This is not the only possibility. The rule $A \Rightarrow C$ can serve also to generalize a query involving term A, under the same assumptions than before. Other possibilities, a formalization of the approach and experiments can be found in [13].

We have also applied FAR's to a web mining problem. Details can be found in [14].

5   Other Patterns

In this section we briefly overview some other text mining paradigms.

5.1   Trends

Trend discovery in text [25] is an application of the mining sequential patterns paradigm [2] to text data. This paradigm is an extension of the associations between itemsets in T-sets. In this case, transactions are labelled with time stamps and other identifiers. A sequence is an ordered list of itemsets. Sequential patterns are sets of items ordered according to their timestamps. The objective is to find maximal sequences (i.e. not included in any other) holding in a set of labelled transactions.

In text, sequences are composed typically of lists of words. Timestamps are obtained from the position of the word in the text. In addition to [2], a method to discover all maximal frequent sequences in text has been proposed in [4].

5.2   Events

In the context of news, event detection tries to discover text data containing previously unknown events. The starting point is an ordered series of news stories.

There are different approaches to solve this problem. One of them is based on clustering documents in terms of their similarity. To calculate the latter, the vector space model is usually employed. Documents that are not similar to prototypes for existing clusters are then candidates to be events. Some works on this area are [48,6].

5.3 Others

Other kinds of patterns and applications in text mining are related to: ontologies [27] and related structures [29], clustering of information extraction frames [22] and prototypical document extraction [36]. As noted in the introduction, our intention is not to be exhaustive (something very difficult in a rapidly growing area), but to show an overview, as wide as possible, of ongoing work in text mining.

6 Discussion

The patterns we can find in data, a representation of real world, are closely related to the knowledge representation paradigm employed. Since they are designed, structured databases reflect a given conceptualization of the world, and that suggests using certain well-known paradigms when looking for patterns in it.

In constrast, textual data presents a very rich and complex inherent structure, consisting of many different features and relations, but we need to make it explicit somehow in order to discover patterns. Many different structured representations of the inherent structure, linked to certain kinds of patterns, have been proposed and employed for text mining purposes. We have provided a non-exhaustive and brief review of some of them.

Despite the actual usefulness of current approaches, there is still a big amount of potential research on this topic, and more complex representations are to be faced by text miners.

Among the important questions that arise in the literature in relation to text structuring, we can mention the determination of the relevance of text features. Research in information retrieval over many years has shown that to measure and to use this relevance improves the performance of systems. In the specific framework of association rules, the paradigms of weighted and fuzzy association rules seem to be suitable for this purpose. We have described our approach to the latter in section 4.

With respect to patterns, the definition of suitable measures of pattern accuracy and importance, according to their semantics, is an important research avenue, specially when adapting existing data mining techniques to textual data. This is the case for example of association rules, where possible different semantics and measures are being studied. In this work we have proposed to use a new framework to assess AR's, that we recently introduced.

References

1. R. Agrawal, T. Imielinski, and A. Swami. Mining association rules between sets of items in large databases. In *Proc. Of the 1993 ACM SIGMOD Conference*, pages 207–216, 1993.
2. R. Agrawal and R. Srikant. Mining sequential patterns. In *Proc. 11th Int. Conf. On Data Engineering*, pages 3–14, 1995.
3. H. Ahonen, O. Heinonen, M. Klemettinen, and A. Inkeri-Verkamo. Applying data mining techniques in text analysis. Technical Report C-1997-23, Department of Computer Science, University of Helsinki, 1997.
4. H. Ahonen-Myka. Finding all frequent maximal sequences in text. In D. Mladenic and M. Grobelnik, editors, *Proc. 16th Int. Conf. On Machine Learning ICML-99 Workshop on Machine Learning in TExt DAta Analysis*, pages 11–17, 1999.
5. H. Ahonen-Myka, O. Heinonen, M. Klemettinen, and A. Inkeri-Verkamo. Finding co-occurring text phrases by combining sequence and frequent set discovery. In R. Feldman, editor, *Proc. 16th Int. Joint Conference on Artificial Intelligence IJCAI-99 Workshop on Text Mining: Foundations, Techniques and Applications*, pages 1–9, 1999.
6. J. Allan, R. Papka, and V. Lavrenko. On-line new event detection and tracking. In *Proc. 21st Annual Int. ACM SIGIR Conf. On Research and Development in Information Retrieval*, 1998.
7. F. Berzal, I. Blanco, D. Sánchez, and M.A. Vila. A new framework to assess association rules. In F. Hoffmann, D.J. Hand, N. Adams, D. Fisher, and G. Guimaraes, editors, *Advances in Intelligent Data Analysis. Fourth International Symposium, IDA'01. Lecture Notes in Computer Science 2189*, pages 95–104. Springer-Verlag, 2001.
8. F. Berzal, I. Blanco, D. Sánchez, and M.A. Vila. Measuring the accuracy and interest of association rules: A new framework. An extension of [7]. Intelligent Data Analysis, submitted, 2002.
9. I. Blanco, M.J. Martín-Bautista, D. Sánchez, and M.A. Vila. On the support of dependencies in relational databases: strong approximate dependencies. Data Mining and Knowledge Discovery, Submitted, 2000.
10. G. Bordogna, P. Carrara, and G. Pasi. Fuzzy approaches to extend boolean information retrieval. In P. Bosc and J. Kacprzyk, editors, *Fuzziness in Database Management Systems*, pages 231–274. Physica-Verlag, 1995.
11. S. Brin, R. Motwani, J.D. Ullman, and S. Tsur. Dynamic itemset counting and implication rules for market basket data. *SIGMOD Record*, 26(2):255–264, 1997.
12. M. Delgado, N. Marín, D. Sánchez, and M.A. Vila. Fuzzy association rules: General model and applications. *IEEE Transactions on Fuzzy Systems*, 2001. Submitted.
13. M. Delgado, M.J. Martín-Bautista, D. Sánchez, J.M. Serrano, and M.A. Vila. Association rules extraction for text mining. FQAS'2002, Submitted, 2002.
14. M. Delgado, M.J. Martín-Bautista, D. Sánchez, J.M. Serrano, and M.A. Vila. Web mining via fuzzy association rules. NAFIPS'2002, Submitted, 2002.
15. M. Delgado, M.J. Martín-Bautista, D. Sánchez, and M.A. Vila. Mining strong approximate dependencies from relational databases. In *Proceedings of IPMU'2000*, 2000.
16. M. Delgado, D. Sánchez, and M.A. Vila. Fuzzy cardinality based evaluation of quantified sentences. *International Journal of Approximate Reasoning*, 23:23–66, 2000.

17. R. Feldman and I. Dagan. Knowledge discovery in textual databases (KDT). In *Proceedings of the 1st Int. Conference on Knowledge Discovery and Data Mining (KDD-95)*, pages 112–117. AAAI Press, 1995.
18. R. Feldman, I. Dagan, and W. Kloegsen. Efficient algorithm for mining and manipulating associations in texts. In *Proc. 13th European Meeting on Cybernetics and Research*, 1996.
19. R. Feldman, M. Fresko, Y. Kinar, Y. Lindell, O. Liphstat, M. Rajman, Y. Schler, and O. Zamir. Text mining at the term level. In *Proc. 2nd European Symposium on Principles of Data Mining and Knowledge Discovery*, pages 65–73, 1998.
20. R. Feldman and H. Hirsh. Mining associations in text in presence of background knowledge. In *Proc 2nd Int. Conf. On Knowledge Discovery and Data Mining, KDD'96*, pages 343–346, 1996.
21. M.A. Hearst. Untangling text data mining. In *Proceedings of the 37 Annual Meeting of the Association for Computational Linguistics*, pages 20–26, 1999.
22. H. Karanikas, C. Tjortjis, and B. Theodoulidis. An approach to text mining using information extraction. In *Proc. Knowledge Management Theory Applications Workshop, (KMTA 2000)*, 2000.
23. Y. Kodratoff. Comparing machine learning and knowledge discovery in DataBases: An application to knowledge discovery in texts. In G. Paliouras, V. Karkaletsis, and C.D. Spyropoulos, editors, *Machine Learning and Its Applications, Advanced Lectures. Lecture Notes in Computer Science Series 2049*, pages 1–21. Springer, 2001.
24. D.H. Kraft and D.A. Buell. Fuzzy sets and generalized boolean retrieval systems. In D. Dubois and H. Prade, editors, *Readings in Fuzzy Sets for Intelligent Systems*, pages 648–659. Morgan Kaufmann Publishers, San Mateo, CA, 1993.
25. B. Lent, R. Agrawal, and R. Srikant. Discovering trends in text databases. In *Proc. 3rd Int. Conference on Knowledge Discovery and Data Mining (KDD-97)*, pages 227–230, 1997.
26. S.-H. Lin, C.-S. Shih, M.C. Chen, J.-M. Ho, M.-T. Ko, and Y.-M. Huang. Extracting classification knowledge of internet documents with mining term associations: A semantic approach. In *Proc. ACM/SIGIR'98*, pages 241–249, 1998.
27. A. Maedche and S. Staab. Mining ontologies from text. In *Proc. 12th International Workshop on Knowledge Engineering and Knowledge Management (EKAW'2000)*, pages 189–202, 2000.
28. H. Mannila and H. Toivonen. Discovering generalized episodes using minimal occurrences. In *Proc. 2nd Int. Conf on Knowledge Discovery and Data Mining (KDD'96)*, pages 146–151, 1996.
29. C.D. Manning. Automatic acquisition of a large subcategorization dictionary from corpora. In *Proc. 31st Annual Meeting of the Association for Computational Linguistics*, pages 235–242, 1993.
30. D. Mladenic. Feature subset selection in text-learning. In *Proc. 10th European Conference on Machine Learning ECML98*, 1998.
31. D. Mladenic and M. Grobelnik. Feature selection for classification based on text hierarchy. In *Working Notes of Learning from Text and the Web, Conference on Automated Learning and Discovery CONALD-98*, 1998.
32. U.Y. Nahm and R.J. Mooney. Using information extraction to aid the discovery of prediction rules from text. In *Proceedings 6th Int. Conference on Knowledge Discovery and Data Mining (KDD-2000) Workshop on Text Mining*, pages 51–58, 2000.
33. U.Y. Nahm and R.J. Mooney. Mining soft-matching rules from textual data. In *Proc. 7th Int. Joint Conference on Artificial Intelligence (IJCAI-01)*, 2001.

34. H.J. Peat and P. Willett. The limitations of term co-occurence data for query expansion in document retrieval systems. *JASIS*, 42(5):378–383, 1991.
35. G. Piatetsky-Shapiro. Discovery, analysis, and presentation of strong rules. In G. Piatetsky-Shapiro and W. Frawley, editors, *Knowledge Discovery in Databases*, pages 229–238. AAAI/MIT Press, 1991.
36. M. Rajman and R. Besançon. Text mining: Natural language techniques and text mining applications. In *Proc. Of the 7th IFIP Working Conference on Database Semantics (DS-7)*. Chapam & Hall, 1997.
37. G. Salton and C. Buckley. Term-weighting approaches in automatic text retrieval. *Information Processing & Management*, 24(5):513–523, 1988.
38. R.C. Schank. Identification of conceptualizations underlying natural language. In R.C. Schank & K.M. Colby, editor, *Compputer Models of Thought and Language*. Freeman, San Francisco, 1973.
39. R.C. Schank. Language and memory. *Cognitive Science*, 4, 1980.
40. E. Shortliffe and B. Buchanan. A model of inexact reasoning in medicine. *Mathematical Biosciences*, 23:351–379, 1975.
41. C. Silverstein, S. Brin, and R. Motwani. Beyond market baskets: Generalizing association rules to dependence rules. *Data Mining and Knowledge Discovery*, 2:39–68, 1998.
42. R. Srikant and R. Agrawal. Mining generalized association rules. In *Proc 21th Int'l Conf. Very Large Data Bases*, pages 407–419, September 1995.
43. Ah-Hwee Tan. Text mining: The state of the art and the challenges. In *Proceedings PAKDD'99 Workshop on Knowledge Discovery from Advanced Databases (KDAD'99)*, pages 71–76, 1999.
44. E.M Voorhees. Query expansion using lexical-semantic relations. In *Proceedings of the 17th Annual Int. ACM/SIGIR Conference*, pages 61–69, 1994.
45. W. Wang, J. Yang, and P.S. Yu. Efficient mining of weighted association rules. In *Proc. Sixth ACM SIGKDD International Conference on Knowledge Discovery & Data Mining*, 2000.
46. K. Winkler and M. Spiliopoulou. Extraction of semantic XML DTDs from texts using data mining techniques. In *Proc. K-CAP 2001 Workshop Knowledge Markup & Semantic Annotation*, 2001.
47. J. Xu and W.B. Croft. Query expansion using local and global document analysis. In *Proceedings 19th Annual Int. ACM/SIGIR Conference on REsearch and Development in Information Retrieval*, pages 4–11, 1996.
48. Y. Yang, T. Pierce, and J. Carbonell. A study on restrospective and online event detection. In *Proc. 21st Annual Int. ACM SIGIR Conf. On Research and Development in Information Retrieval*, pages 28–36, 1998.
49. L. A. Zadeh. A computational approach to fuzzy quantifiers in natural languages. *Computing and Mathematics with Applications*, 9(1):149–184, 1983.

Modelling and Incorporating Background Knowledge in the Web Mining Process

Myra Spiliopoulou and Carsten Pohle

Department of E-Business, Leipzig Graduate School of Management
{myra,cpohle}@ebusiness.hhl.de

Abstract. The incorporation and exploitation of background knowledge in KDD is essential for the effective discovery of useful patterns and the elimination of trivial results. We observe background knowledge as a combination of beliefs and interestingness measures. In conventional data mining, background knowledge refers to the preferences and properties of the population under observation. In applications analysing the interaction of persons with a system, we identify one additional type of background knowledge, namely about the *strategies* encountered in pursuit of the interaction objectives. We propose a framework for the modelling of this type of background knowledge and use a template-based mining language to exploit it during the data mining process. We apply our framework on Web usage mining for Web marketing applications.

1 Introduction

Background knowledge is essential for all applications where knowledge discovery with data mining techniques is used. Part of the domain expert's knowledge is exploited during the data preparation phase: It encompasses well-accepted facts about the properties of the data in the domain, such as that pregnancy is only encountered among female persons, but also advances of domain research. In fact, background knowledge is formed through personal observation, interaction with other people and familiarity with the research advances in the application domain.

In related research, we encounter two *implicit* types of background knowledge: (i) beliefs of experts, usually modelled as interrelated rules and (ii) interestingness criteria, usually modelled as statistic functions applied over the discovered patterns [18,22,23,6]. We can combine those two approaches in saying that *an expert has some beliefs about the population delivering the data and is interested in patterns that verify, contradict or modify her beliefs.* Interestingness is then a function quantifying the verification, contradiction or modification of a belief according to the data, whereby the belief may or may not be stated explicitly.

In application domains related to interactive environments, we encounter a further type of background knowledge, namely the one related to the interaction of the object under observation with its environment. The difference between the three types is best expressed by an example:

Example 1 *In market-basket analysis, one may encounter the rule that people buying nachos often purchase some dip as well. Whether this rule is interesting or not depends*

D.J. Hand et al. (Eds.): Pattern Detection and Discovery, LNAI 2447, pp. 154–169, 2002.

on the beliefs of the marketing expert about the preferences of supermarket customers. To exploit this rule for the placement of products in the supermarket, the marketing expert needs one more piece of knowledge, namely how customers move inside the supermarket. Her background knowledge might tell her that some customers have a list of products to buy and will stick to them unless they encounter a closely related product nearby. Her background knowledge might also tell her that other customers just walk about the supermarket to find the week's specials or new products. This is a different kind of background knowledge than the one related to customer preferences. It refers to the behavioural "strategies" of the customers inside the supermarket.

Hence, when studying the interactivity aspects of an application, a further important type of background knowledge emerges, namely knowledge about the *strategy* for pursuing the objective of the interaction. A strategy is formed and followed by the interacting person and *not* by the designer of the environment, where this interaction takes place. However, the designer (in general: the application expert) has some background knowledge on what such a strategy will be like.

In this study, we investigate the modelling and exploitation of background knowledge for Web usage analysis. Background knowledge in this domain concerns (a) the preferences of the users to the contents of the site and (b) the strategy with which the users pursue their objectives inside the site. Consider for example a site selling books. Aspect (a) refers to beliefs and discovered rules of the form "if object A, then object B", where A and B can be books, subjects, authors or delivery options. Aspect (b) refers to whether users invoke the search engine or rather browse across links, whether they consider the properties of many products before purchasing one or rather buy whatever product they come across. We concentrate here on the second aspect, which is as important as the first one, because it implies different site designs and different types of recommendations for each group of users.

We propose a mechanism for the modelling of background knowledge during data preparation and for the mapping of beliefs about user behaviour into templates of a mining language. Our contribution lays (i) in the identification, modelling and exploitation of an important type of background knowledge that is not addressed in the past and (ii) in the specification of a procedure for the incorporation of background knowledge into the mining process. We would like to stress that our framework is not peculiar to Web usage analysis: The analysis of Web user behaviour is our application field, selected for two reasons. First, this application domain demonstrates that both user preferences and user interaction with a Web site are essential to understand the users and make recommendations to them. Second, our approach exploits the background knowledge of a domain expert, so that we chose a domain where we can bring the necessary expertise. As will be made apparent from our definition of interaction strategy, other domains where interaction is expressed as a sequence of tasks can also profit from our approach. Such are behavioural analysis, intrusion detection and workflow management.

In the next section, we introduce our knowledge incorporation framework by describing its components and suggesting a procedure for integrating them into the data preparation and the data mining process. In section 3, we apply our framework to model shopping strategies, as proposed in marketing theory, and to test them against Web usage data. Section 4 describes related work. The last section concludes our study.

2   A Knowledge Incorproration Framework

In our knowledge incorporation framework, background knowledge consists of:

(i)   the *tasks* expected to be performed by users of the interactive environment
(ii) the *utilities* available to accomplish such tasks
(iii) combinations of utility invocations in pursuit of a goal; these are intended *strategies* of the users
(iv) *templates* of strategies allowing for the data-driven generation and evaluation of strategies upon the patterns discovered by the miner.

Central to our framework is the concept of "strategy". In strategic management, Mintzberg et al [16], say that "strategy is a pattern, that is, consistency in behavior over time". They contrast "strategy as a plan", when referring to the future plans of an organization, with "strategy as a pattern", when referring to the past behaviour of the organization. The first is termed "intended strategy", the latter is the "realized strategy" [16].

Although persons are not organizations, we anticipate that the above definitions can be applied in our context of the interaction between individuals and their environment: The intended strategy is in the mind of the system user. The application expert can observe only the tasks and utilities invoked by the user, i.e. the realized strategy as *a pattern extracted by the miner* from the data recorded by the system.

In most application domains, the expert attempts to interpret the realized strategy and assess the intended one from it according to some theory. Depending on the domain, the theory may come from strategic management, Web marketing, advances in network intrusion detection or medicine for chronic diseases. We concentrate on theories from Web marketing to assess intended strategies of users in B2C Web-commerce.

2.1   Tasks

In the proposed framework, tasks are individual activities performed to achieve a goal feasible inside the interaction environment. Tasks in a Web site may be the retrieval of a product description, the submission of an order or the playback of an audio sequence.

The specification of an application's task is beyond the scope of our framework, which rather suggests how they should be modelled: As classes of objects, as states in state diagrams or as components of business processes. At the same time, the categorization of tasks at different levels of abstraction must be supported, e.g. by concept hierarchies [20], class inheritance diagrams in UML or functional diagrams displaying tasks and subtasks in business process modelling. For the analysis of the data pertaining to user interactions, the specification of tasks must adhere to the high-level concepts of the application and not to the objects of the software tool performing the interaction.

Example 2 *For some database applications, it may be important to distinguish between queries with aggregate operators and queries without, and further state the presence or absence of subqueries.*

In market-basket analysis, the application may require the specification of product categories or the categorization of information material according to its depth and confidentiality level. If the analysis is applied on an e-shop, the invocations of Web site objects like URLs, scripts and images should be mapped onto tasks like the placement of a product into the shopping basket.

2.2   Utilities for Tasks Execution

A task may be realized by invoking different utilities. For example, a list of products may be acquired by querying a search engine or clicking into a product catalogue; a product description may adhere to a number of presentation settings. Occasionally, the utility enclosing a task cannot be distinguished from the task itself: If a product can be presented concisely and in full detail, should one model this as two utilities upon the same task or as two tasks invoked by the same utility?

In [4], we do not distinguish utilities from tasks, but rather categorize tasks across two *complementary* concept hierarchies, a "content-driven" and a "service-driven" one. The former type captures the contents of the tasks in the application's context. The latter type refers to the utilities being invoked to render the tasks.

In our knowledge incorporation framework, we suggest the same separation of concept hierarchy types and use the term "task" to cover both the task and the utility rendering it. Such a combination can be realized in the data preparation phase by expressing the properties of tasks according to multiple dimensions: Content-driven concept hierarchies can be found in most institutions that maintain a data warehouse. If the application considers the rendering of the tasks by different utilities as essential, tasks can be modelled additionally across a service-driven concept hierarchy.

Example 3 *In Web merchandizing, the first high-level task accomplished by potential customers is that of information acquisition [12]. We can refine this generic specification of a task in two ways, either with respect to the type of information being acquired or with respect to the service used for information acquisition.*

The first aspect can be realized in a content-driven concept hierarchy. An example taxonomy of this type for the Web site used in our experiments is shown in Fig. 1.

The second aspect can be realized by means of service-driven concept hierarchies: The expert may distinguish between the invocation of a query processor and the navigation across arbitrary links, thus modelling the distinction between people that prefer to search and those that prefer to navigate [21]. The former concept can be further refined with respect to the number of properties used in each query or the number of queries issued before a product inspection. In [4], we describe a set of service-driven concept hierarchies upon a Web-site for searchers and show how the analysis of navigational behaviour revealed shortcomings in the design of the search interface.

2.3   Strategies as Sequences of Tasks

We define a *strategy* as a sequence of tasks, beginning at a *start-task* and ending at a *target-task* that corresponds to the fullfillment of the objective of the interaction. In-between may appear an arbitrary number of intermediate tasks, whereby some of them may be performed more than once.

More formally, let the set of conceivable tasks in an application be a set of symbols S. Then, a strategy is a regular expression involving at least two symbols from S (the start-task and the target-task) and, optionally, a number of "iterators". We propose the following notation and constraints[1]:

- An iterator has the form $[n; m]$, where n is a non-negative integer, m is a non-negative integer or a symbol denoting infinity, and $n \leq m$.
- An iterator $[n; m]$ appears as suffix to a task or a parenthesized subsequence of symbols. It stands for at least n and at most m occurrences.
- There is a dummy task *ANY*, which may appear inside the strategy with or without an iterator attached to it. It can be matched by any task in S.
- The dummy task *ANY* cannot be the first nor the last task of a strategy.
- The first task or subsequence of a strategy may be prefixed by the symbol #, stating that interactions conforming to the strategy start with this very task.

The reader may notice that while tasks have been defined in subsection 2.1 with reference to powerful modelling tools like UML, the definition of a strategy allows for much less expressiveness as provided e.g. by UML statechart diagrams or by Petri nets. The reason is that a strategy, as we conceive it in our knowledge incorporation framework, should be matched against *recorded* sequences of tasks from a dataset like a Web server log: While more sophisticated representations are conceivable, any of them should be mapped into a plain sequence of events by some translator software. Our proposed notion of strategy is powerful enough to express background knowledge of application-experts and is appropriate for direct analysis, using a powerful mining language. Moreover, it can serve as target representation for a translator designed for a more powerful representation, such as a Petri net.

Example 4 *According to [12], a potential customer undergoes three phases until the successful purchase, namely (i) information acquisition, (ii) negotiation and (iii) purchase. Web-merchandizing sites are designed to serve visitors appropriately during each of these phases. Cutler and Sterne state that a good site should be sticky in the information acquisition phase and slippery afterwards [9].*

When products are sold electronically, information acquisition concerns the retrieval and inspection of the products of potential interest. The purchase corresponds to filling and submitting an order. The negotiation may include the selection of the most appropriate shipping option or the specification of discounts, upon which a customer has claim. In an e-auction, the negotiation phase is the phase of bidding.

Obviously, the aforementioned phases do not necessarily occur once for a given customer. On the contrary, it is very natural and quite desirable (from the merchandizer's point of view) that a visitor who placed a product in her electronic basket inspects further products. This may lead to reconsideration of the shipping options and potentially to higher discounts. A higher discount may motivate the inspection of more products. Hence, the tasks of acquiring information on a product and negotiating upon the contents of a cart may be repeated several times.

---

[1] The notation adheres to the syntax of the mining language MINT [25], which we use for strategy evaluation, as will be shown in the sequel.

In our notation, the theory of [12] suggests the following strategy:

$$\#(InformationAcquisition[1; \infty]Negotiation)[1; \infty]Purchase,$$

allowing for at least one acquisition of information before performing a negotiation. This subsequence may be repeated an arbitrary number of times before a purchase takes place. The # symbol at the beginning indicates that this strategy is only matched by sessions starting at the task "InformationAcquisition": A session starting at the task "JobOfferings" would not match the above strategy, even if it contained the tasks of "InformationAcquisition", "Negotiation" and "Purchase" later on. Such sessions would be covered by the more general strategy

$$(InformationAcquisition[1; \infty]Negotiation)[1; \infty]Purchase.$$

The first strategy requires that the users do nothing beyond acquiring information and negotiating before the purchase. The second strategy allows them to perform other tasks, too, before entering the phase of information acquisition. Whether the theory of [12] is closer to the first or the second strategy is beyond the scope of this paper.

These two strategies subsume the simpler strategy below, which permits only one negotiation task to be performed directly before the purchase.

$$\#InformationAcquisition[1; \infty]Negotiation\ Purchase$$

In our example, we have mapped the activities of users in a Web-merchandizing site into a set of strategies, one generalizing the other, and derived from a well-established theory in the application domain. Hence, the expert can start her analysis by exploiting renown theories instead of trying to invent a new one. She can cross-check the strategies derived from the theory, verify whether the most restrictive or rather a more generic strategy describes most of the data, or whether there are many sessions not adhering to the derived strategies at all. In the latter case, the expert recognises that the theory indeed does not fit (all of) the data and is called to test for the existence of strategies that are not yet reported in the domain theory.

The modelling of strategies as sequences of tasks and iterators raises the following challenges for the data analysis:

1. What is the support of each strategy observed in a usage log?
2. How do we instruct a miner to extract the pattern(s) corresponding to an interaction strategy?
3. If strategies overlap, which is the earliest task at which we can uniquely identify each strategy and thus predict the behaviour of the user thereafter?

The first question is reflected in the statistics of the pattern(s) corresponding to a strategy. The second issue is dealt with by formulating templates of strategies in a mining language. We observe the third issue as future work. However, in our experiment we show how this case can be dealt with in a concrete application.

2.4   A Strategy as an Interaction Pattern

According to subsection 2.3, a strategy is a sequence of tasks, allowing for an arbitrary number of repetitions of a task or a task subsequence. In the log describing the interaction sessions between a user and a system, a session supports a strategy, if it contains the tasks required by the strategy, repeated as many times as allowed by the repetition operators in the strategy description. Accordingly, a strategy is a sequential pattern, whose support is computed from the number of sessions containing this pattern.

This approach, although semantically correct, is not satisfactory. Consider the simple marketing strategy $\#InformationAcquisition[1; \infty]Negotiation\,Purchase$ described in Example 4 above, according to which users acquire information for one or more products and do a purchase after negotiating only once. The support of this pattern, i.e. the number of customers that purchase according to this strategy is obviously of interest. However, the number of users that visited the site according to this strategy and gave it up midway is not of less interest: These users should be assisted or motivated to follow their goal to the end. This latter number is reflected by the support of each task in its position inside the strategy.

In general, let s be a strategy composed of tasks and iterators, and let $u = a_1 \ldots a_n$ be a usage sequence (in general: part of a session) adhering to it. We adorn each task a_i, $i = 1, \ldots, n$, with the number of sessions containing $a_1 \ldots a_i$, divided by all sessions. This is the support of the task in the sequence pattern of this strategy:

$$support(a_i, u) = \frac{card(\{x \in L, x = y \cdot a_1 \ldots a_i \cdot z\})}{card(L)} \tag{1}$$

where L is the usage log and $card(\cdot)$ computes the cardinality of a collection that may contain duplicates. Then, for each task in the strategy, we can derive the confidence with which it leads to the next task:

$$conf(a_{i+1}, a_i, u) = \frac{support(a_{i+1}, u)}{support(a_i, u)} \tag{2}$$

A strategy defined on the basis of tasks and wildcards matches more than one session in a usage log. For example, a strategy $s_0 = a[1; 3]b$ is supported by $u_1 = ab$, $u_2 = aab$ and $u_3 = aaab$. Then, the support of the first a for the strategy s_0 is $support(a, s_0) = support(a, u_1) + support(a, u_2) + support(a, u_3)$. The support of the second a is $support(aa, s_0) = support(aa, u_2) + support(aa, u_3)$. The support of b is $support(b, s_0) = support(b, u_1) + support(b, u_2) + support(b, u_3)$. To express these statistics, we need to aggregate all sequences supporting the strategy into a single pattern, which is no more a sequence but a graph.

In [24], the notion of "navigation pattern" has been proposed with semantics that are appropriate for this purpose. Informally, a navigation pattern describes a g-sequence. A g-sequence is a sequence comprised of events (here: tasks) and optional wildcards between them. For a g-sequence $e_1 * e_2 * \ldots * e_n$ of events $e_1, \ldots, e_n$ interleaved with wildcards, the navigation pattern is an array of n trees. The root of the ith tree ($i = 1, \ldots, n - 1$) is e_i; its branches are all the subsequences in the log that lead from e_i to e_{i+1} and are preceded by $e_1 * \ldots * e_{i-1}$. The nth tree consists of e_n only.

Example 5 *Let $L = a, ab, dacb, bc$. Let $a * b$ be a g-sequence. Its navigation pattern consists of two trees. The second one is the node b, whose support is 2. The first tree has a as its root, with support 3, and two branches ab and acb. Each node of the tree is adorned with the number of sequences leading to this node. This corresponds to the notion of support in Eq. 1.*

For a support threshold of 50%, the g-sequence $a * b$ is frequent, being supported by the sequences ab and dacb. This means that the events a, b in this g-sequence appear frequently in that order. The branches of the first tree are not frequent. They mark the paths connecting the frequent events and are skipped by most miners. For our modelling of strategies, the paths are essential.

We use the notion of navigation pattern to aggregate the usage sessions that support a strategy. Let a strategy s consist of n tasks $\tau_1, \ldots, \tau_n$ with iterators between some of them. We map this strategy into the g-sequence $\tau_1 * \tau_2 * \tau_n$. Thereby, we constrain the wildcard after τ_i ($i = 1, \ldots, n - 1$):

- The only event that can match the wildcard is τ_i itself.
- If the iterator after τ_i has the form $[k_i; m_i]$, then number of events matched by the wildcard is at least $k_i - 1$ and at most $m_i - 1$.

Hence, a strategy $s = a[1; 3]b$ is mapped into the g-sequence $a * b$, where the wildcard after a can only be matched by zero to 2 *further* appearances of a.

By mapping strategies to navigation patterns of g-sequences, we achieve two things: First, all information pertinent to a strategy in terms of contents and statistics is captured by the data stored in a navigation pattern. Second, the evaluation of anticipated interaction strategies is mapped into the discovery and analysis of the (navigation) patterns corresponding to these strategies.

The proposed mapping scheme has a number of shortcomings: First, the aforementioned notion of pattern is only supported by the Web usage miner WUM [26], so that conventional miners cannot be used for strategy evaluation. Second, the pattern discovery process relies on the expressiveness of the mining language MINT [25], which is used to perform the mapping. Currently, we concentrate on allowing for strategy evaluation by *any means*. The realization of the process by general purpose software is observed as future research.

2.5 Mining Queries for the Evaluation of Strategies

For the discovery of the strategies upon the usage data, we use the Web usage miner WUM [26]. This miner discovers navigation patterns. As shown in the end of the previous subsection, a strategy can be expressed in such a way that it corresponds to a navigation pattern subject to some constraints. These constraints are expressed in the template-based mining language MINT that guides the miner. A template is a sequence of variables and wildcards. The variables are bound to events, while the wildcards are bound to subsessions connecting these events. In the context of strategy discovery, the variables range over the tasks in the set of symbols S. The wildcards should simulate the iterators. This is achieved by restricting the range of a wildcard similarly to the range of the iterator and by restricting the content of the wildcard to the task associated with the iterator.

Example 6 *Recall the marketing strategy* #*InformationAcquisition*[1; ∞]*Negotiation Purchase described in Example 4. The set of symbols S contains only three tasks, InformationAcquisition, Negotiation, Purchase. In the language* **MINT***, this strategy is expressed as:*

```
SELECT t
FROM NODE AS x, y, z, TEMPLATE t AS # x * y z
WHERE x.url = "InformationAcquisition"
AND y.url = "Negotiation"
AND z.url = "Purchase"
AND wildcard.y.url = "InformationAcquisition"
```

The template expresses the strategy as a sequence of variables and wildcards. The first three constraints bind the variables. The last constraint binds the contents of the wildcard. The range of the wildcard is from 0 to infinity, so it does not need to be bound.

The result of the example mining query is a single navigation pattern, since all variables are bound. The template-based language, though, allows for more, e.g. for the analysis of incomplete strategies and of their most frequent points of exit.

Example 7 *In Web-merchandizing, the attrition during the visit in the Web site is a serious phaenomenon. The query in Example 6 returns the support and detailed statistics of the strategy* #*InformationAcquisition*[1; ∞]*Negotiation Purchase. It is essential to know where users go if they abondon this strategy after the negotation phase.*

```
SELECT t
FROM NODE AS x, y, z, TEMPLATE t AS # x * y z
WHERE x.url = "InformationAcquisition"
AND y.url = "Negotiation"
AND wildcard.y.url = "InformationAcquisition"
AND ( z.support / y.support ) > 0.01
```

This query has the variable z unbound. It only places a statistical constraint upon the confidence of z (cf. Eq. 2). The miner will return one pattern per value of z, i.e. per exit from this incomplete strategy. Depending on the confidence, with which the negotation leads to a purchase, the complete strategy may also be one of the resulting patterns.

MINT allows for the specification of strategies and of bundles of similar strategies, so that more than one patterns are returned and can be compared. However, **MINT** does not currently support the whole expressiveness needed to specify a template for any strategy. For example, the nesting of iterators is not supported.

Example 8 *The general strategy in Example 4 contains nested iterators.*

$$\#(InformationAcquisition[1; \infty]Negotiation)[1; \infty]Purchase$$

We exploit the fact that the nested iteration involves two of the total of three tasks.

```
SELECT t
FROM NODE AS x, y, z, TEMPLATE t AS # x * y * z
WHERE x.url = "InformationAcquisition"
AND y.url = "Negotiation"
AND wildcard.y.url = "InformationAcquisition"
AND wildcard.z.url != "Purchase"
AND z.url = "Purchase"
```

Alternatively, we could exploit the string similarity between InformationAcquisition and Negotiation; such desirable similarities can be enforced during data preparation.

```
AND wildcard.z.url ENDSWITH "tion"
```

The two alternative constraints are equivalent for this set of tasks. They return again a single pattern, because all variables are bound. This pattern is not completely equivalent to the strategy above, as it allows for consecutive invocations of the Negotiation task. Since a negotiation is a task performed once, unless a further product is considered (and information is acquired for it), the extra subpattern may not exist at all.

3   A Case Study on Web Site Serving as a Point-of-Contact

In the following, we apply our knowledge incorporation framework to test interaction strategies upon the traffic data of a non-merchandizing site. Companies that provide sophisticated services asking for customisation or a long-term investment (e.g. ASPs) use the Web mainly as a contact point, in which visitors can be informed about the company, its product portfolio, its credentials and references. The objective of the site is then to motivate visitors to seek a face-to-face contact.

In our analysis, we have focussed on potential customers and have removed entries identified as belonging to actual customers or personnel, robots, archivers and administration services. Sessions were determined heuristically (cf. [8,4,3]), specifying 30 minutes as a threshold for viewing a single page of a session. After cleaning and preprocessing, the cleaned server log contained 27647 sessions.

3.1   Strategies for Different Information Demands

Although the site of our experiment is not a Web-merchandizing site, we consider it appropriate to investigate the applicability of marketing theories on the behaviour of visitors upon it. We opt for the model of Moe [17]: In her study on the behaviour of in-store customers, Moe distinguishes four shopping strategies, differing in the pages being visited and in the duration of each access [17]. Of interest for our case site is the strategy of *knowledge-building*. Knowledge-builders seek a complete picture of the institution; the impression they get may influence their future purchasing behaviour. In our case site, we investigate to what extent this strategy leads to contact establishment.

Figure 1 shows the content-oriented taxonomy and the task-oriented taxonomy of the Web site. The reader may notice that the latter contains most objects of the former but in a different arrangement. For example, "DetailInfo" is a subconcept below "Marketing/PR" in the content-based taxonomy, but also a sibling task to *"BackgroundInfo"*, which generalizes the task "Marketing/PR" in the task-based hierarchy. The tasks in italics are those having no counterpart in the content-oriented taxonomy; they generalize objects common to both taxonomies.

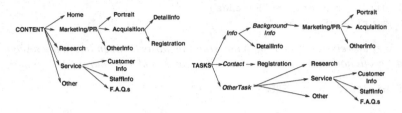

Fig. 1. Content- and task-based taxonomies of the Web site

According to the task-oriented taxonomy of Fig. 1, the knowledge building strategy has the form:

$$\# Home\ (BackgroundInfo[1;n]\ DetailInfo[1;n])[1;n] \qquad (3)$$

Visitors adhering to this strategy will acquire background information before asking for detailed information. After invoking "DetailInfo" for the first time, knowledge builders may opt for a contact immediately or they may ask for further background or detail information. People desiring an overview of the institution may ask for both types of information. Hence, we abstracted them into the concept "Info" of the task-oriented hierarchy in Fig. 1. We transform this strategy into the following MINT query:

```
SELECT t
FROM NODE AS x y z w, TEMPLATE # x y * w * z AS t
WHERE x.url = "Home" AND y.url = "BackgroundInfo"
AND wildcard.w.url = "BackgroundInfo"
AND w.url = "DetailInfo"
AND wildcard.z.url ENDSWITH "Info" AND z.url = "Contact"
```

3.2   Analysis of the Discovered Patterns

The knowledge building strategy of Eq. 3 corresponds to the navigation pattern returned by the MINT query depicted in the end of the previous section. As explained in section 2.4, a navigation pattern consists of a group of paths representing alternative routes [24]. In our case, these are the routes from "BackgroundInfo" to "DetailInfo" and then to the invocation of the "Contact" task. All these paths consist of "BackgroundInfo" and "DetailInfo" tasks. However, one visitor may have invoked "DetailInfo" after asking for "BackgroundInfo" once, while another may have requested "BackgroundInfo" 10 times beforehand. Moreover, each path has been entered by a number of visitors, some of which have followed it to the end, while others have abandoned it.

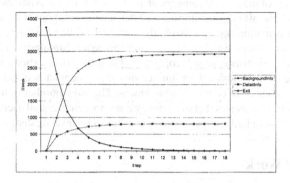

Fig. 2. First component: Knowledge-building strategy until the first invocation of "DetailInfo"

The invocation of detailed information indicates a serious interest in the offered product or service. Hence, we split the pattern of this strategy into two components, one until the first invocation of "DetailInfo" and one thereafter. The statistics of the first component of the knowledge-building strategy are shown in Fig. 2. The horizontal axis represents steps, i.e. task invocations. At each step, a number of users asks for detailed information and thus proceeds to the second component of the strategy. These users are represented in the cumulative curve labelled "DetailInfo". The vertical axis shows that from the circa 3800 visitors that entered this strategy, about 21% (800 visitors) entered the second component. The remaining ones are depicted in the cumulative curve labelled "Exit": They did not necessarily bandon the site, but their subsequent behaviour does not correspond to the knowledge-building strategy any more. The curve labelled "BackgroundInfo", represents the visitors that ask for further background information. All curves saturate fast, i.e. most users invoke only a few tasks.

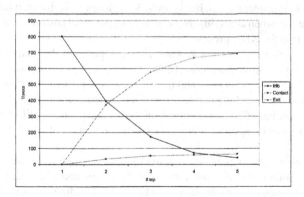

Fig. 3. Second component: Knowledge-building strategy until contact establishment

The statistics of the second component are shown in Fig. 3. After the first invocation of "DetailInfo", the 800 visitors that entered the second component acquired detailed or background information aggregated into the curve "Info" that covers invocations of both tasks. The "Contact"-curve and the "Exit"-curve are again cumulative. The former shows that 10% of these visitors establish contact, and that they do so after a small number of information acquisition tasks. This implies that many contact acquisition tasks do not increase the confidence of contact establishment. The large number of users represented by the "Exit"-curve indicates that the strategy does not represent all users. Hence, further tasks should be modelled and more strategies should be investigated.

4 Related Work

Our knowledge incorporation framework stands in the junction of several domains, where background knowledge has been addressed, though in a different context.

Beliefs and Interesting Patterns. The notion of belief appears in several domains. Beliefs about user behaviour in the Web have been used to identify unexpected and actionable patterns [22,18]. However, such beliefs are association rules or sequences, not interaction strategies as suggested by researchers in the application domain.

Several frameworks have been proposed to identify interesting patterns, these being defined as unexpected or surprising ones with respect to the expert's beliefs. The notion of interestingness for association rules and sequences has been considered, among others in [19,11,22,27]; surprising rules for Web applications are studied in [1]. Such approaches concentrate on the statistics of discovered rules. The structural and content aspects of an interaction strategy enjoy less interest.

Languages and cost models for Web usage mining. The Web as new interaction medium challenges designers and enterpreneurs, who seek the best strategies to present content and gett revenue from Web sites. Pattern discovery can be performed with conventional techniques like market basket analysis and frequent sequence mining. However, the information about user behaviour registered by a server is much richer than the data registered in off-line interactions. Background knowledge of Web experts and application experts is valuable when analysing the behaviour of potential customers, and thus should be integrated into the KDD process.

Dedicated Web usage mining systems are equipped with languages that allow for sophisticated expert-miner interaction [2,25]. Other researchers map user behaviour into scalar values like conversion rates and their derivatives [5,13,26,9] or into properties of the site like stickiness and slipperiness [9]. Such measures, although valuable, capture the average behaviour of all users: They cannot distinguish among different behavioural strategies that focus on specific contents or on the invocation of specific services.

Data preparation. The mapping of system objects onto application objects is essential for the exploitation of beliefs, interestingness functions and conversion rates. System objects can be mapped onto application objects by building concept hierarchies, as is done in the OLAP cube of a data warehouse. In [20], we argue that the concept hierarchies being established as part of the data warehouse are appropriate for regularly invoked OLAP queries but not for ad hoc analysis, nor for the generation and testing of hypotheses. In [4], we introduced the distinction between content-driven and "service-driven" concept hierarchies. The latter describe the services being invoked in pursue of a business objective, and correspond to rendering "utilities" in our terminology.

The steps of the KDD process. The components of the KDD process could be modelled as tasks in a workflow management system. Efforts in this direction include the specification of the KDD modules and their combinations into a KDD process, as performed in CRISP-DM [7], the XML-based specification PMML for data mining [10] and the data mining specification for OLE [15].

Our framework is complementary to such approaches. We concentrate on the data preparation and data analysis components. For these, we consider quite sophisticated activities, like the modelling of strategies, the combination of multiple concept hierarchies, and the specification of mining templates over strategies.

5 Conclusions

We have presented a framework for the incorporation of background knowledge in the mining process, focussing on knowledge in the form of anticipated behaviour strategies in interactive environments. We model strategies as sequences of recurrent tasks. We show how the contents and the rendering aspects of such tasks can be captured, how tasks can be combined to strategies and how strategies can be expressed in a mining language to guide the pattern discovery process. We have used our framework to model interaction strategies in Web marketing and reported on the results for a Web site.

The procedure of incorporating background knowledge according to our framework is not yet a formal set of activities with well-defined interfaces. We intend to investigate the functionalities offered by models like CRISP-DM and by XML specifications like PMML, and attempt to express the knowlege incorporation procedure in such a model.

Our model of interactive strategies is appropriate for domains where interaction can be expressed as a sequence of tasks. However, our current realization is limited by the capabilities of the sequence miner WUM, on top of which our mining language MINT is implemented.Hence, we investigate the possibilities of applying MINT constructs to further data mining algorithms.

Interaction strategies may overlap, in the sense that a subsequence of tasks may belong to more than one strategy. Dynamically adjustable interactive systems and recommendation engines rely on the identification of the user's interaction strategy at the earliest timepoint possible. Hence, we plan to investigate how the characteristic tasks or task sequences of overlapping interaction strategies can be detected.

References

1. G. Adomavicius and A. Tuzhilin. Discovery of actionable patterns in databases: The action hierarchy approach. In *KDD*, pages 111–114, Newport Beach, CA, Aug. 1997.

2. M. Baumgarten, A. G. Büchner, S. S. Anand, M. D. Mulvenna, and J. G. Hughes. Navigation pattern discovery from internet data. In *Proceedings volume [14]*, pages 70–87. 2000.
3. B. Berendt, B. Mobasher, M. Spiliopoulou, and J. Wiltshire. Measuring the accuracy of sessionizers for web usage analysis. In *Int. SIAM Workshop on Web Mining*, Apr 2001.
4. B. Berendt and M. Spiliopoulou. Analysing navigation behaviour in web sites integrating multiple information systems. *VLDB Journal, Special Issue on Databases and the Web*, 9(1):56–75, 2000.
5. P. Berthon, L. F. Pitt, and R. T. Watson. The world wide web as an advertising medium. *Journal of Advertising Research*, 36(1):43–54, 1996.
6. S. Chakrabarti, S. Sarawagi, and B. Dom. Mining surprising patterns using temporal description length. In A. Gupta, O. Shmueli, and J. Widom, editors, *VLDB'98*, pages 606–617, New York City, NY, Aug. 1998. Morgan Kaufmann.
7. P. e. a. Chapman. *CRISP-DM 1.0. A Step-by-Step Data Mining Guide*. The CRISP-DM Consortium, www.crisp-dm.org, 2000.
8. R. Cooley, B. Mobasher, and J. Srivastava. Data preparation for mining world wide web browsing patterns. *Journal of Knowledge and Information Systems*, 1(1), 1999.
9. M. Cutler and J. Sterne. E-metrics — business metrics for the new economy. Technical report, NetGenesis Corp., http://www.netgen.com/emetrics, 2000. access date: July 22, 2001.
10. DMG, www.dmg.org/pmmlspecs_v2/pmml_v20.html. *An Introduction to PMML 2.0.*
11. A. Freitas. On objective measures of rule surprisingness. In *PKDD'98*, number 1510 in LNAI, pages 1–9, Nantes, France, Sep. 1998. Springer-Verlag.
12. B. Ives and G. Learmoth. The information system as a competitive weapon. *Communications of the ACM*, 27(12):1193–1201, 1984.
13. J. Lee, M. Podlaseck, E. Schonberg, R. Hoch, and S. Gomory. Analysis and visualization of metrics for online merchandizing. In *Proceedings volume [14]*, pages 123–138. 2000.
14. B. Masand and M. Spiliopoulou, editors. *Advances in Web Usage Mining and User Profiling: Proceedings of the WEBKDD'99 Workshop*, LNAI 1836. Springer Verlag, July 2000.
15. Microsoft Corporation, www.microsoft.com/data/oledb/dm.htm. *OLE DB for Data Mining Specification 1.0*, July 2000.
16. H. Mintzberg, B. Ahlstrand, and J. Lampel. *Strategy Safari: A Guided Tour through the Wilds of Strategic Management*. THE FREE PRESS, a division of Simon & Schuster Inc., New York, 1998.
17. W. W. Moe. Buying, searching, or browsing: Differentiating between online shoppers using in-store navigational clickstream. *Journal of Consumer Psychology*, (forthcoming), 2001.
18. B. Padmanabhan and A. Tuzhilin. A belief-driven method for discovering unexpected patterns. In *KDD'98*, pages 94–100, New York City, NY, Aug. 1998.
19. G. Piateski-Shapiro and C. J. Matheus. The interestingness of deviations. In *AAAI'94 Workshop Knowledge Discocery in Databases*, pages 25–36. AAAI Press, 1994.
20. C. Pohle and M. Spiliopoulou. Building and exploiting ad hoc concept hierarchies for web log analysis. In *Proc. of 4th Int. Conf. on Data Warehousing and Knowledge Discovery, DaWaK 2002*, Aix-en-Prevence, Sept. 2002. to appear.
21. B. Shneiderman. Designing information-abundant web sites: Issues and recommendations. *JHCS; Extracted and adapted from "Designing the User Interface: Strategies for Effective Human-Computer Interaction", 3rd edn., Reading MA: Addison-Wesley, 1998*, 1997. www.hbuk.co.uk/ap/ijhcs/webusability/shneiderman/shneiderman.html.
22. A. Silberschatz and A. Tuzhilin. What makes patterns interesting in knowledge discovery systems. *IEEE Trans. on Knowledge and Data Eng.*, 8(6):970–974, Dec. 1996.
23. P. Smyth and R. Goodmann. An information theoretic approach to rule induction from databases. *IEEE Trans. on Knowledge and Data Eng.*, 4(4):301–316, 1992.
24. M. Spiliopoulou. The laborious way from data mining to web mining. *Int. Journal of Comp. Sys., Sci. & Eng., Special Issue on "Semantics of the Web"*, 14:113–126, Mar. 1999.

25. M. Spiliopoulou and L. C. Faulstich. WUM: A Tool for Web Utilization Analysis. In *extended version of Proc. EDBT Workshop WebDB'98*, LNCS 1590, pages 184–203. Springer Verlag, 1999.

26. M. Spiliopoulou and C. Pohle. Data mining for measuring and improving the success of web sites. In R. Kohavi and F. Provost, editors, *Journal of Data Mining and Knowledge Discovery, Special Issue on E-commerce*, volume 5, pages 85–114. Kluwer Academic Publishers, Jan.-Apr. 2001.

27. K. Wang, W. Tay, and B. Liu. Interestingness-based interval merger for numeric association rules. In *4th Int. Conf. of Knowledge Discovery and Data Mining KDD'98*, New York City, NY, Aug. 1998. AAAI Press.

Modeling Information in Textual Data Combining Labeled and Unlabeled Data

Dunja Mladenić

J.Stefan Institute, Ljubljana, Slovenia and
Carnegie Mellon University, Pittsburgh, USA
Dunja.Mladenic@{ijs.si,cs.cmu.edu},
http://www-ai.ijs.si/DunjaMladenic/

Abstract. The paper describes two approaches to modeling word normalization (such as replacing "wrote" or "writing" by "write") based
on the re-occurring patterns in: word suffix and the context of word
obtained from texts. In order to collect patterns, we first represent the
data using two independent feature sets and then find the patterns responsible for a particular word mapping. The modeling is based on a set
of hand-labeled words of the form (word, normalized word) and texts
from 28 novels obtained from the Web and used to get words context.
Since the hand-labeling is a demanding task we investigate the possibility of improving our modeling by gradually adding unlabeled examples.
Namely, we use the initial model based on word suffix to predict the
labels. Then we enlarge the training set by the examples with predicted
labels for which the model is the most certain. The experiment show that
this helps the context-based approach while largely hurting the suffix-
based approach. To get an idea of the influence of the number of labeled
instead of unlabeled examples, we give a comparison with the situation
when simply more labeled data is provided.

1 Introduction

Modeling information in text data is becoming more popular with the large
amount of texts available in electronic form and especially on the Web. A commonly addressed problem is document categorization that enables performing
tasks such as e-mail filtering or document ontology maintenance. The work described here addresses the problem of finding reoccurring patterns inside words
or in the word context in order to group words that have the same normalized
form. A similar approach is word stemming where words having the same stem
are considered as members of the same group of words. For stemming of English words, there are known algorithms giving approximation of the mapping
from a word to its stem, eg. Porter's stemmer [1]. However, stemming words in
other natural languages requires separate consideration. In particular, there is
no publicly available algorithm for stemming Slovenian words and the same is
true for many other languages. The additional difficulty of our problem is that
we are dealing with highly inflected natural language, having up to 30 different

D.J. Hand et al. (Eds.): Pattern Detection and Discovery, LNAI 2447, pp. 170–179, 2002.

word forms for the same normalized word. If we want to apply some of the existing approaches for handling text data based on word frequencies to natural languages having many different forms for the same word, we should first apply some kind of word normalization to get a reasonable underlying statistics.

The related research on learning English past tense addresses a subproblem of a general word normalization problem addressed here. Some researchers use decision trees or neural networks to predict separate letters of the transformed word based on the letters in the past tense form of the word [2]. Some work using relational learning with decision lists was also proposed for learning English past tense [3] and learning lemmatization of Slovenian words [4]. The latter addresses similar problem on the same natural language as addressed here, but requires having information from part of speech tagger (that is not in general available for Slovenian and thus a part of speech tagger has to be trained first) and handles only some types of the words. The main difference is that we aim at word normalization as pre-processing for other Text Learning tasks potentially involving large number of different words. Thus we want to be general by handling any type of word and texts not necessary provided in the usual form of sentences. For instance, in the Yahoo! hierarchy of Web documents each document is described by a one-line summary that can be view as a training example [5]. Moreover in our work described here we use a list of words. There is also some related work incorporating context but in unsupervised learning capturing different morphological transformations [6] that among others uses a natural language parser and some hand-constructed rules for defining the context of a word. In our work the context is automatically captured from a fixed size window or words surrounding the word under consideration.

The addressed problem has some special properties compared to the related work in text classification. We are classifying a single word and not larger amount of text as usual in text classification. Section 2 gives motivation and description of the problem. Data is described in Section 3 followed by the approach description in Section 4. Experimental results are given in Section 5 and discussed in Section 6.

2   Problem Definition

The problem addressed in this paper is learning a mapping of words based on some labeled data defining the mapping and a set of texts in the same natural language. The mapping is from many to one, where we are mapping from up to 30 different forms of the same word to the normalized form of the word. We illustrate the approach on the problem of word normalization and define a mapping in an indirect way as a transformation on a word suffix. This transformation is from different forms of the same word to the normalized form of the word. In our problem presentation, the words with the same class value are the words having the same suffix transformation and not the words that have the same normalized form. For instance, *working, laughing, eating* would be all assigned the class value mapping from *ing* to an empty string. Notice that in the original problem *working* is in the same class with *works, worked, work*, as the words with the normalized form *work*. We can say that the problem definition used in this work is based on having class values reflecting the common suffix transformation of the words, while in a different setting the common normalized form could be used instead. Using both problem formulations as two complement views to the same problem, and combining their results is potential part of future work.

Our motivation here is to be able to apply the learned mapping on new text, and output the transformed text. One of the direct applications of such system is in using it as pre-processor of text for document categorization or search engines. This is especially valuable for texts written in highly inflected natural language such as Slovenian, Czeck or Croatian, where an average word has over 20 different forms.

3   Data Description

Our labeled data sets were obtained as a random sample of about 2000 examples taken from a hand-constructed lexicon [7]. The whole lexicon contains about 20 000 different normalized words with listed different forms for each of them resulting in about 500 000 different entries (potential learning examples). For each word its normalized form is provided and additionally, the word is annotated by some information about the word form such as singular /plural, noun/verb/adjective, etc. Since our motivation is to develop an approach that can be used on any texts in the selected natural language and this additional information is not readily available in the usual texts, we are not using it here. However, one can argue that if a language tagger is available and the text is in the appropriate form to apply it, this additional information can be very valuable.

The context was obtained from a set of 28 novels (4Mb of text) known from Slovenian literature. Since our set of words taken from the labeled data set contains all different inflections of the included normalized words, only about 15% of these inflected words were found in the used independent set of documents. For the remaining words, we could not generate representation based on the word context. Thus the experiments using the context features were performed only on these 15% of the words. Testing the approach using much larger set of unlabeled documents is part of the future work. Namely, we plan to use a subset of Slovenian corpus containing about 40 000 Web documents obtained in experiments for "automatic Web search query generation to create minority language corpora" [8].

4   Approach Description

Learning a specific transformation of a word that is needed to obtain its normalized form requires representing the word in a suitable form. We identified two independent representations based on the: (1) suffix of a word itself and (2) context of a word from some set of documents. In this paper we describe experiments using these two representations independently and combining them using a variant of Co-training [9]. By applying co-training we take advantage of unlabeled data by predicting the labels, selecting a small subset that has the best predictions and adding this subset with "hard" labels on examples to the original labeled data set. Notice that this approach differs from the related approach where EM algorithm is used to probabilistically label unlabeled examples, as for instance used for document categorization [10]. For experimental comparison showing that on the selected data set co-training outperforms EM see [11].

For co-training we have used (1) letter-based representation to provide labels for unlabeled data. In each co-training step up to 50 unlabeled examples are labeled with the predicted class value and added to the training set. Namely, the trained letter-based model is applied on all unlabeled examples. Then we select examples for prediction of which the model is the most certain about and add them together with the predicted label. After that both models are retrained on the enlarged training set and tested using the fixed testing set. The testing set is formed in the cross-validation on the set of labeled examples.

4.1   Letter-Based Representation

We define the letter-based representation of word as a set of features giving word suffix of a fixed length. In our experiments, we used up to five last letters as a suffix. In that way, each word is represented with five discrete features. For examples, English word *laughing* is represented in the following form: LastCh = "g", LastTwoCh = "ng", LastThreeCh = "ing", LastFourCh = "hing", LastFiveCh = "ghing". Distribution of the number of examples over different values of our attributes shows that most examples are group around the same feature values, leaving many under supported values. For instance, for the attribute showing the last letter of the word (word suffix of size 1), top 7 attribute values contain over 85% of examples. These values and their corresponding number of examples are the following: A 1116 examples, I 679, E 603, O 528, M 469, H 268, U 180.

The problem of mapping different word forms to its normalized form was defined as a data mining problem on which we applied standard classification rules for finding the mapping patterns. The algorithm we used in our experiments was developed in ATRIS rule induction shell [12]. The algorithm is based on local optimization and adopts the covering paradigm, giving results similar to other symbolic learning algorithms.

4.2   Context-Based Representation

We defined the context of a word to contain up to six words in a window around the word and the word itself. Each word is thus represented by a set of words collected from all the contexts that the word appears in. As described in Section 3 only about 15% of the labeled words were found in the texts used for word context.

The problem here was defined as text mining problem on which we applied a Naive Bayesian classifier base on the multinomial event model as outlined in [13].

The algorithm has been shown to work well on a number of text classification problems despite the used word independence assumption.

Context-based word representation can be very short, containing just a couple of words, but this depends on the goodness of the match between the labeled and unlabeled data. Namely, this will not be the case if the unlabeled data contains several occurrences of the labeled words. Nevertheless, using the Naive Bayesian classifier on relatively short documents is problematic, since there is not enough evidence for classifying testing examples. In order to preserve information about inflection, we use all the words without changing them (no stop-list or stemming in applied). In many cases, very few or none of the words from a testing example is found in the training examples, and when the word is found in the training examples it may be distributed over examples having different class values. The relatively low number of different words for each class value contributes to unreliable probability estimates used in the classification.

4.3   Lables of Examples

The class value giving a label to example was defined by giving the transformation to be applied on the word in order to get the normalized form. The transformation is given in the form of map X to Y (XToY), where X is a suffix of the word and Y is a suffix of the normalized word. For example, taking English word *laughing* would give a mapping to "laugh" where its class value would be given as a transformation of deleting the ending "ing" and we write that as the class value "ingTo_". Since our class value is strongly connected to the last letter of a word, we applied a simple way of *Sequential modeling* [14]. Namely, instead of taking all the training examples with 486 different class values, we generated a one level decision tree (decision stump) that divided the examples into subsets depending on the last letter in the word. In this way we decomposed the problem into 20 independent subproblems, each having between 1 and 40 class values (most of the subproblems have less than 10 class values) instead of 486 class values. Because of the nature of the problem, the words that have different last letter can share the class value only if that value is "do not change the word" (in our notation value _To_) or "add some suffix" (_ToY). This means that with the applied sequential modeling the problem was divided into subproblems with almost non-overlapping class values. Here the already described procedure for co-training is applied on each subproblem.

When observing the distribution of examples over 486 class values on one of the cross-validation splits, we noticed that the 20 most common class values contain about 45% of the examples. The most common class values and the corresponding number of examples (in the set of 4580 training examples) are as follows: _To_: 317 examples, MATo_ 115, IHTo_ 111, _ToI 105, AToI 102, EToI 100, LIToTI 100, LOToTI 99, LAToTI 98, LToTI 94, LEToTI 91, MITo_ 91, HTo_ 91. The decomposition of a problem according to the last word letter (here referred to as sequential modeling) resulted in 20 subproblems (one for each ending letter of a word) and again, only a small number of class values contain the majority of examples. For instance, the training set for a subproblem of

words ending with letter "m" has 469 training examples, 45 different class values
where only 23 of them have more than 1 example. The most frequent class values
covering 84% of training examples and the corresponding number of examples
(on one of the cross-validation splits) for that subproblem are the following:
MTo_ 111 examples, MToTI 67, IMTo_ 60, EMTo_ 50, OMTo_ 38, EMToI 25,
NEMToEN 23, NIMToEN 20. Sequential modeling reduced the number of class
values inside each subproblem and improved the classification results, as shown in
Section 5. The classification performance on the letter-based word representation
is improved as well as on the context-based word representation for 10% and 20%
respectively.

5   Experimental Results

Experiments were performed on two sets of independent features: (1) letters
in a word and (2) context of a word (see Section 4). We report the results of
classification accuracy from 5-fold cross-validation, where we used exactly the
same examples, but represented in two different ways. We used subsets of the
whole labeled data set split randomly into 5 folds. The whole training-testing
loop was thus repeated five times, with each of the five subsets being excluded
from training and used as testing exactly once (cross-validation). We performed
experiments on a samples having 1818 examples representing 1% of the whole
labeled data set. As can be seen in Figure 1 the performance varies with the
number of training examples.

Our definition of class value as a transformation of the word suffix resulted in
some class values having a very small number of examples, that is additionally
reduced by employing cross-validation. In our experiments using the letter-based
representation, such small classes were ignored as noise meaning that no rules
were generated for them.

We have experimentally tested the following hypothesis. **Hypothesis 1**: including more labeled examples improves performance of both approaches. **Hypothesis 2**: including unlabeled examples labeling them using the letter-based
approach improves performance of the context-based approach. The results of
experiments are summarized in Table 2.

We have **confirmed the first hypothesis** finding that both approaches
perform significantly better using all 1800 training examples than using the initial set of 450 training examples. This difference is not significant though when
the context-based approach is used in the sequential modeling framework. **The
second hypothesis was partially confirmed** showing improvement of the
context-based approach where the classification accuracy increased from 10% to
23% on the original problem and from 40% to 44% for the sequential modeling (see Figure 2). At the same time performance of the letter-based approach
largely drops from 42% to 20% on the original problem and from 49% to 31% for
the sequential modeling . Similar regularity was found on a larger data sample
having in total 5725 examples, where we added up to 200 unlabeled examples
in one co-training step. Starting with 1375 training examples and adding up to

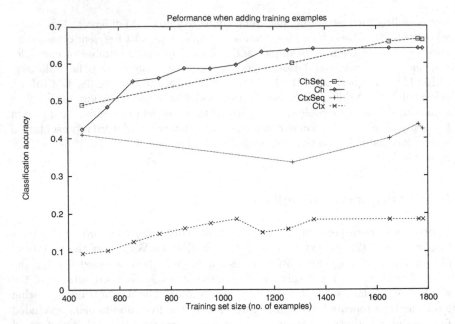

Fig. 1. Influence of the used number of training examples on the performance of two different approaches: letter-based (Ch) and context-based (Ctx) approach. Both are used on the original problem and in the framework of sequential modeling (ChSeq, CtxSeq). It can be seen that including more training examples improve performance on the testing examples for all but the last approach (CtxSq). It is also evident that the sequential modeling improves performance of both approaches.

1800 unlabeled examples resulted in decrease of the letter-based approach performance from 60% to 41% and small increase in the context-based approach the performance from 16% to 18%. We expect that the results would be improved when using not only a larger initial training set but also more conservative addition of unlabeled examples (adding only 50 or 100 instead of 200 unlabeled examples in each co-training step).

6   Discussion

The problem addressed in this paper is finding reoccurring patterns inside words (sequences of letters) or in the word context (sets of words) in order to group words that have the same normalized form. Consequently, we identified two independent representations based on the suffix of a word itself and the context of a word from some set of documents. We describe experiments using these two representations independently and combining them using a variant of co-training based on unlabeled data. Namely, the hand-labeling of data is a demanding task thus we investigated the possibility of improving our modeling by gradually

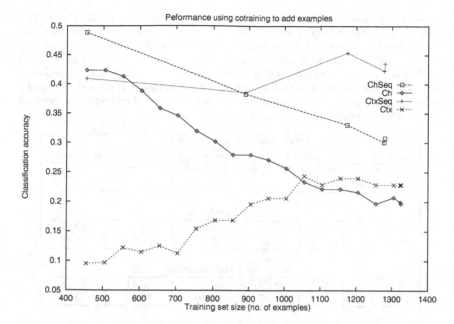

Fig. 2. Influence of adding unlabeled examples on the performance of different approaches to word normalization. Notice that we use only 450 labeled training examples and gradually add up to 800 unlabeled examples.

adding unlabeled examples. The idea is in generating the initial model from a small set of labeled examples and using it for assigning labels to unlabeled examples that are usually easy to obtain. In our case unlabeled data consists simply of words from the natural language we are addressing. We used the letter-based representation to predict the labels of unlabeled examples, since that representation significantly outperforms the context-based representation. The experiment show that adding unlabeled examples in this way helps the context-based approach while largely hurts the letter-based approach. We expect that some of the problems we have observed with the context-based approach will be solved by using a larger set of documents containing more words from our labeled data.

Sequential modeling was used to obtain a set of simpler problems. These new problems mostly have up to 10 class values, instead of 486 class values that we have in the original problem. Our experiments show that this resulted in an improvement of classification accuracy of 10% and 20% on the two independent example representations we used, which thus achieved classification accuracy of 72.8% and 40.8% respectively. To get an idea about the overall performance, we have obtained results of an existing stemmer incorporating hand-coded rules [15] on our data. We found that the stemmer in 79% of cases grouped the same words together as provided in our hand-labeled data for word lemmatization (we can not tell if the stem is correct or not, thus we have checked the overlap between the

Table 1. Experimental comparison of co-training influence on the letter-based (Ch) and the context-based (Ctx) word representation used for transforming the word into its normalized form. For each representation we also show the performance when using the sequential modeling to divide the problem into subproblems. We report average classification accuracy and standard deviation on 5-fold cross validation experiments. The reported results are on the sample having 1818 examples taking 450 examples as labeled and the rest as unlabeled. We also show the influence of adding more, randomly selected labeled examples.

Classification accuracy [%]				
Approach	Co-training using Ch predictions			Adding labeled examples
no. of exs (train+unlab.)	450 + 0	450 + 500	450 + 800	1300	1800
		Original problem			
Ch	42.36	27.06	19.68	63.10	63.97
Ctx	9.52	20.59	22.87	15.95	18.46
		Using sequential modeling			
ChSeq	48.86	38.32	30.76	60.11	66.26
CtxSeq	40.94	38.57	43.56	33.56	42.43

proposed and the correct word grouping). As for future work, we see a possibility of cooperation with a language expert to get feedback on the patterns found.

References

[1] M.F. Porter. An algorithm for suffix stripping. In *In ACM SIGIR Conference on Research and Development in Information Retrieval*, pages 318–327, 1980.

[2] C. X. Ling. Learning the past tense of English verbs: The symbolic pattern associator vs. connectionist models. *Journal of Artificial Intelligence Research*, 1:209–229, 1994.

[3] R.J. Mooney and M.E. Califf. Induction of first-order decision lists: Results on learning the past tense of english verbs. In L. De Raedt, editor, *Proceedings of the 5th International Workshop on Inductive Logic Programming*, pages 145–146. Department of Computer Science, Katholieke Universiteit Leuven, 1995.

[4] Saso Dzeroski and Tomaz Erjavec. Learning to lemmatise slovene words. In *Learning language in logic, (Lecture notes in computer science, J.Cussens and S.Dzeroski (eds))*, pages 69–88, 2000.

[5] Dunja Mladenic and Marko Grobelnik. Feature selection for unbalanced class distribution and naive bayes. In *Proceedings of the 16th International Conference on Machine Learning*, 1999.

[6] David Yarowsky. Unsupervised word sense disambiguation rivaling supervised methods. In *Meeting of the Association for Computational Linguistics*, pages 189–196, 1995.

[7] Tomaz Erjavec. The multext-east slovene lexicon. In *Proceedings of the 7th Slovene Electrotechnical Conference ERK-98*, 1998.

[8] Rayid Ghani, Rosie Jones, and Dunja Mladenic. Automatic web search query generation to create minority language corpora. In *Proceedings of the Sixteenth Annual International ACM SIGIR Conference on Research and Development in Information Retrieval*, 2001.

[9] Blum and Mitchell. Combining labeled and unlabeled data with co-training. In *COLT: Proceedings of the Workshop on Computational Learning Theory, Morgan Kaufmann Publishers*, 1998.

[10] Kamal Nigam, Andrew McCallum, Sebastian Thrun, and Tom Mitchell. Text classification from labeled and unlabeled documents using em. *Machine Learning*, 39(2/3):103–134, 2000.

[11] Kamal Nigam and Rayid Ghani. Analyzing the effectiveness and applicability of co-training. In *Ninth International Conference on Information and Knowledge Management*, 2000.

[12] Dunja Mladenic. Combinatorial optimization in inductive concept learning. In *Proc. 10th Int. Conf. on Machine Learning, Morgan Kaufmann*, pages 205–211, 1993.

[13] A. McCallum and K. Nigam. A comparison of event models for naive bayes text classifiers. In *AAAI-98 Workshop on Learning for Text Categorization*, 1998.

[14] Yair Even-Zohar and Dan Roth. A sequential model for multi-class classification. In *Proc. of Conference on Empirical Methods in Natural Language Processing (EMNLP 2001)*, 2001.

[15] Jure Dimec, Saso Dzeroski, Ljupco Todorovski, and Dimitrij Hristovski. Www search engine for slovenian and english medical documents. In *Stud Health Technol Inform.:68*, 1999.

Discovery of Frequent Word Sequences in Text

Helena Ahonen-Myka

University of Helsinki
Department of Computer Science
P.O. Box 26 (Teollisuuskatu 23)
FIN–00014 University of Helsinki, Finland,
helena.ahonen-myka@cs.helsinki.fi

Abstract. We have developed a method that extracts all maximal frequent word sequences from the documents of a collection. A sequence is said to be frequent if it appears in more than σ documents, in which σ is the frequency threshold given. Furthermore, a sequence is maximal, if no other frequent sequence exists that contains this sequence. The words of a sequence do not have to appear in text consecutively.
In this paper, we describe briefly the method for finding all maximal frequent word sequences in text and then extend the method for extracting generalized sequences from annotated texts, where each word has a set of additional, e.g. morphological, features attached to it. We aim at discovering patterns which preserve as many features as possible such that the frequency of the pattern still exceeds the frequency threshold given.

1 Introduction

We have developed an automatic method for discovering textual patterns that can be used as compact content descriptors of documents [1,2,3]. The patterns have the form of a word sequence, i.e., we attempt to extract from the text a small set of word sequences that describe the contents of the document. Word sequences were chosen as a representation, since they have great potential to be a rich computational representation for documents, such that, on the one hand, feature sets for various further forms of analysis (e.g. text classification) can be easily retrieved, but, on the other hand, also a human-readable description of the document (e.g. a summary) can be generated from the representation.

Our discovery method extracts all the maximal frequent word sequences from the text. A sequence is said to be frequent if it appears in more than σ documents, in which σ is the frequency threshold given. Furthermore, a sequence is maximal, if no other frequent sequence exists that contains this sequence. The words of a sequence do not have to appear in text consecutively: a parameter g tells how many other words two words in a sequence can have between them. The parameter g usually gets values $1-3$. For instance, if $g = 2$, in both of the following two text fragments:

...President of the United States Bush...
...President George Bush...

D.J. Hand et al. (Eds.): Pattern Detection and Discovery, LNAI 2447, pp. 180–189, 2002.

a sequence *president bush* would be found (the articles and prepositions are not counted as words).

The ability to extract maximal sequences of any length, i.e., also very long sequences, and the allowance of gaps between words of the sequence distinguish our method from the other methods that could be used for extracting word sequences from text, e.g. the work on sequential patterns by Agrawal and Srikant [4] and the work on episodes by Mannila, Toivonen, and Verkamo [5]. The gaps make the word sequences flexible: the usual variety in natural language can be addressed. In addition to that long maximal sequences are very descriptive, they are also very compact representations. If we restricted the length of the sequences to, e.g., 8 words, and there actually were a frequent sequence of 25 words in the collection, we would find thousands of sequences that only represent the same knowledge as the one maximal sequence.

In this paper, we describe briefly the method for finding all maximal frequent word sequences in text and then extend the method for extracting generalized sequences from annotated texts, where each word has a set of additional, e.g. morphological, features attached to it. We aim at discovering patterns which preserve as many features as possible such that the frequency of the pattern still exceeds the frequency threshold given.

The following application illustrates the problem. A common type of language analysis is to create a *concordance* for some word, i.e., to list all the occurrences of the word in a text corpus, with some left and right context. A human analyst can then study the contexts and try to gather some generalized knowledge about the use of the word. If there are many occurrences, it may not be easy to characterize the contexts, particularly if the word has several senses. For instance, consider the following occurrences of a word *'right'*:

Is that the   right time?
...that things weren't right between us.
Stay   right here.
They had the   right to strike.

In characterizing the contexts, the analyst might find out that some other words seem to occur frequently together with the word, or that the surrounding words belong to some class of words (e.g. nouns), or that they have a special form (e.g. singular, accusative). We can easily obtain this kind of information for words by using morphological analysis. A morphological analysis attaches each word with, e.g., the base form, number, case, tense, and part of speech. This process is rather fast with current tools. The analysis of a concordance could be automated by discovering generalized sequences, in which the infrequent features are removed, whereas features that appear often — and are probably more significant for the structures studied — are preserved, e.g.:

the   right 'Noun'
be   right between 'Pronoun'

'Verb' right here
the   right to 'Verb'

The paper is organized as follows. In Section 2 the method for discovering maximal frequent word sequences is described. In Section 3 the method is extended to find generalized sequences in morphologically annotated texts. Section 4 contains some discussion.

2   Finding Maximal Frequent Word Sequences

Assume S is a set of documents, and each document consists of a sequence of words.

Definition 1. *A sequence $p = a_1 \cdots a_k$ is a* subsequence *of a sequence q if all the items $a_i, 1 \leq i \leq k$, occur in q and they occur in the same order as in p. If a sequence p is a subsequence of a sequence q, we also say that p occurs in q.*

Definition 2. *A sequence p is* frequent *in S if p is a subsequence of at least σ documents of S, where σ is a given frequency threshold.*

Note that we only count one occurrence of a sequence in a document: several occurrences within one document do not make the sequence more frequent.

Definition 3. *A sequence p is a* maximal frequent (sub)sequence *in S if there does not exist any sequence p' in S such that p is a subsequence of p' and p' is frequent in S.*

The discovery process is presented in Algorithm 1. In the *initial phase* (steps 1–3) we collect all the ordered pairs, or 2-grams, (A, B) such that words A and B occur in the same document in this order and the pair is frequent in the document collection. Moreover, we restrict the distance of the words of a pair by defining a maximal gap; in our experiments we used a maximal gap of 2, meaning that at most 2 other words may be between the words of a pair.

The maximal frequent sequences are extracted in the *discovery phase* of Algorithm 1 (steps 4–19), which combines bottom-up and greedy approaches. A straightforward bottom-up approach is inefficient, since it would require as many levels as is the length of the longest maximal frequent sequence. When long maximal frequent sequences exist in the collection, this can be prohibitive, since on every level the join operation increases exponentially the number of the grams contained in the maximal frequent sequences. Although the greedy approach increases the workload during the first passes, the gain in efficiency is still substantial.

In the discovery phase, we take a pair and *expand* it by adding items to it, in a greedy manner, until the longer sequence is no more frequent. The occurrences of longer sequences are computed from the occurrences of pairs. All the occurrences computed are stored, i.e., the computation for ABC may help to compute later the frequency for $ABCD$. In the same way, we go through all pairs, but we

Algorithm 1 *Discovery of all maximal frequent subsequences in the document collection.*

Input: *S: a set of documents, σ: a frequency threshold*
Output: *Max: the set of maximal frequent sequences*

// *Initial phase: collect all frequent pairs.*
1. For all the documents $d \in S$
2.   collect all the ordered pairs within d
3. $G_2 =$ all the ordered pairs that are frequent in S

// *Discovery phase: build longer sequences by*
// *expanding and joining grams.*
4. $k := 2$
5. $Max := \emptyset$
6. While G_k is not empty
7.   For all grams $g \in G_k$
8.   If g is not a subsequence of some $m \in Max$
9.   If g is frequent
10. $max := Expand(g)$
11. $Max := Max \cup max$
12.   If $max = g$
13.   Remove g from G_k
14.   Else
15.   Remove g from G_k
16.   Prune(G_k)
17. $G_{k+1} := Join(G_k)$
18. $k := k + 1$
19. Return Max

only try to expand a pair if it is not already a subsequence of some maximal sequence, which guarantees that the same maximal sequence is not discovered several times. When all the pairs have been processed, every pair belongs to some maximal sequence. If some pair cannot be expanded, it is itself a maximal sequence. If we knew that every maximal sequence contains at least one unique pair, which distinguishes the sequence from the other maximal sequence, then one pass through the pairs would discover all the maximal sequences. As this cannot be guaranteed, the process must be repeated iteratively with longer k-grams.

In the expansion step (step 10) of Algorithm 1, all the possibilities to expand have to be checked, i.e., at any point, the new item can be added to the tail, to the front or in the middle of the sequence. If one expansion does not produce a frequent sequence, other alternatives have to be checked. The expansion is greedy, however, since after expanding successfully it proceeds to continue expansion, rather than considers alternatives for the expansion. The choice of items to be inserted is restricted by the k-grams, i.e., also after expansion the sequence is constructed from the existing k-grams.

In the next step, we *join* pairs to form 3-grams, e.g., if there exist pairs AB, BC, and BD, we form new sequences ABC and ABD. Then we make a pass over all these 3-grams, and, as with the pairs, we try to expand grams that are not subsequences of the known maximal sequence and that are frequent. We can always remove those grams that are themselves maximal sequences, since such a gram cannot be contained in any other maximal sequence. The discovery proceeds respectively, variating expansion and join steps, until there are no grams left in the set of grams.

Algorithm 2 *Prune.*
Input: G_k: a gram set
Output: G_k: a pruned gram set

1.   *For each $g = a_1 \cdots a_k \in G_k$*
2.   *Let $LMax_g = \{p \mid p \in Max$ and $a_1 \cdots a_{k-1}$ is a subsequence of $p\}$*
3.   *Let $RMax_g = \{p \mid p \in Max$ and $a_2 \cdots a_k$ is a subsequence of $p\}$*
4.   *For each $p = b_1 \cdots b_n \in LMax_g$*
5.   *$LStr_{p,g} = \{b_1 \cdots b_{i_1-1} \mid a_1 \cdots a_{k-1}$ occurs in $i_1 \cdots i_{k-1}$ in $p\}$*
6.   *For each $p = b_1 \cdots b_n \in RMax_g$*
7.   *$RStr_{p,g} = \{b_{i_k+1} \cdots b_n \mid a_2 \cdots a_k$ occurs in $i_2 \cdots i_k$ in $p\}$*
8.   *$LStr_g = \{LStr_{p,g} \mid p \in LMax_g\}$*
9.   *$RStr_g = \{RStr_{p,g} \mid p \in RMax_g\}$*
10.   *For each $s_1 \in LStr_g$*
11.   *For each $s_2 \in RStr_g$*
12.   *$s_{new} = s_1.g.s_2$*
13.   *If s_{new} is not a subsequence of a maximal sequence*
14.   *For each frequent subsequence s of s_{new}*
15.   *If s is not a subsequence of a maximal sequence*
16.   *Mark all grams of s*
17. *For each $g = a_1 \cdots a_k \in G_i$*
18.   *If g is not marked*
19.   *Remove g from G_k*

Often it is not necessary to wait until the length of the grams is the same as the length of a maximal sequence, in order to remove a gram from the set of grams. After a discovery pass over the set of grams, every gram is a subsequence of at least one maximal sequence. Moreover, any new maximal sequence that can be generated has to contain grams either from at least two maximal sequences or two grams from one maximal sequence in a different order than in the existing maximal sequence. Otherwise a new sequence would be a subsequence of an existing maximal sequence.

This motivates the *pruning* phase of the algorithm, which proceeds as follows. For every gram it is checked how the gram might join existing maximal sequence to form new sequences. If a new candidate sequence is not a subsequence of some existing maximal sequence, all subsequences of the candidate sequence are considered in order to find new frequent sequences that are not contained in any maximal sequence, remembering that if a sequence is not frequent, its supersequences cannot be frequent. If a frequent sequence is found, all its grams are marked. After all grams are processed, grams that are not marked are removed from the gram set.

3   Finding Maximal Frequent Generalized Feature Sequences

We now move on to consider annotated documents, like in the sample collection of four documents in Figure 1. We assume again that S is a set of documents, but now each document consists of a sequence of feature vectors.

```
* Document 1
i       i       nom     pron
saw     see     past    v
a       a       sg      det
red     red     abs     a
ball    ball    nom     n
and     and     nil     cc
a       a       sg      det
green   green   abs     a
ball    ball    nom     n
* Document 2
the     the     nil     det
red     red     abs     a
ball    ball    nom     n
was     be      past    v
small   small   abs     a
* Document 3
the     the     nil     det
green   green   abs     a
ball    ball    nom     n
was     be      past    v
big     big     abs     a
* Document 4
he      he      nom     pron
saw     see     past    v
the     the     nil     det
balls   ball    nom     n
as      as      nil     adv
well    well    nil     adv
```

Fig. 1. A sample document collection.

In the sample, the first feature is an inflected word, the second is the base form, and the fourth is the part of speech. The third feature varies based on the part of speech of the word. For instance, 'nom' means nominative (nouns, pronouns), 'abs' an absolute (adjectives; as opposite to comparatives and superlatives), and 'past' means a past tense (verbs). Some parts of speech (adverbials, determiners) do not have any special information. Thus, the third feature has a value 'nil' for them.

We model each feature vector as an ordered set of feature values. In order to simplify the problem, though, we make some restrictions. First, we need the following notation.

Definition 4. *Let* $r[u_i, \ldots, u_j]$, $1 \leq i \leq k$, $i \leq j \leq k$, *be the set of occurrences of the feature vectors* u *in the document collection for which* $\{u_i, \ldots, u_j\} \subset u$.

The constraints are the following.

- Each feature vector has k feature values, i.e. $u = <u_1, \ldots, u_k>$, and the ith value of each vector, $1 \leq i \leq k$ represents a comparable generalization level within all the vectors.
- Dropping a feature value from the beginning of a feature vector generalizes the feature vector. That is, $r[u_i, \ldots, u_k] \subseteq r[u_j, \ldots, u_k]$, in which $1 \leq i \leq k$, $i \leq j \leq k$.

The second condition holds for many interesting linguistic features, like the ones in our sample. However, this condition does not hold, if we consider at the same time features like *number, case, and gender* for nouns, or *tense, mood and aspect* for verbs, since these features are on the same generalization level.

Note that the second condition could not be replaced by using taxonomies. That is, there is not necessarily an ISA relationship between feature values u_i and $u_{i'}$, $i < i' \leq k$. For instance, from a feature vector $< saw, see, past, v >$ we cannot deduce that all the words with a base form 'see' are in the 'past' tense.

The definitions for word sequences can be updated for the generalized case in an obvious way.

Definition 5. *A sequence* $p = a_1 \cdots a_k$ *is a g-subsequence of a sequence* $q = b_1 \cdots b_n$, *if there exists a sequence of feature vectors* $s = b_{j_1} \cdots b_{j_k}$ *such that* b_{j_i} *occurs before* $b_{j_{i+1}}$ *in* q, $1 \leq i \leq k$, *and* $a_i \subseteq b_{j_i}$ *for each* i.

Definition 6. *A sequence* p *is g-frequent in* S *if* p *is a g-subsequence of at least* σ *documents of* S, *where* σ *is a given frequency threshold.*

Definition 7. *A sequence* p *is a* maximal g-frequent (sub)sequence *in* S *if there does not exist any sequence* p' *in* S *such that* p *is a g-subsequence of* p' *and* p' *is g-frequent in* S.

We can apply Algorithm 1 for the generalized case by modifying the initial phase. In the initial phase, all the frequent ordered pairs are collected. In Algorithm 1, two words that have at most g other words between them are collected, the frequencies of pairs are counted, and the frequent pairs are selected. The main difference, when we consider generalized sequences, is that, even if two feature vectors u and v that occur close enough may not be frequent in the collection, a pair (u', v'), where $u' \subseteq u$ and $v' \subseteq v$, may be. Hence, the initial phase of Algorithm 1 has to be enhanced by discovering which subsets of the feature vectors form frequent pairs.

The straightforward way would be to collect for each pair of feature vectors $(< u_1, \ldots, u_n >, < v_1, \ldots, v_m >)$ all the pairs $(< u_i, \ldots, u_n >, < v_j, \ldots, v_m >)$, in which $1 \leq i \leq n$ and $1 \leq j \leq m$ and count the frequencies. There are, however, some possibilities for optimization. The Lemma 9 below expresses a similar constraint as the 'Apriori trick', which is used in the discovery of association rules [6].

Lemma 1. *If a feature value f does not occur in at least σ documents, any sequence of feature vectors containing f is not g-frequent.*

The following lemma uses the knowledge of the features, as given in the conditions above.

Lemma 2. *Let $u = < u_1, \ldots, u_n >$ and $v = < v_1, \ldots, v_m >$ be feature vectors. If a pair $(< u_i, \ldots, u_n >, < v_j, \ldots, v_m >)$ is not g-frequent, also any pair $(< u_{i'}, \ldots, u_i, \ldots, u_n >, < v_{j'}, \ldots, v_j, \ldots, v_m >)$, in which $1 \leq i' \leq i$ and $1 \leq j' \leq j$, is not g-frequent.*

Based on the lemmas above, in the initial phase, the infrequent feature values are pruned from the feature vectors. Moreover, for each ordered pair of feature vectors, we can first collect all the pairs of suffixes. The frequencies of the pairs of suffixes can then be used to guide the creation of combinations in the following way.

If a pair of suffixes $(< u_i, \ldots, u_n >, < v_j, \ldots, v_m >)$ is frequent, in which $1 \leq i \leq n$ and $1 \leq j \leq m$, all the pairs $(< u_{i''}, \ldots, u_n >, < v_{j''}, \ldots, v_m >)$, in which $i \leq i'' \leq n$ and $j \leq j'' \leq m$, are frequent and, hence, are added to the set of 2-grams. Moreover, the pairs $(< u_i, \ldots, u_n >, < v_{j'''}, \ldots, v_m >)$, in which $1 \leq i \leq n$ and $j \leq j''' \leq m$, and the pairs $(< u_{i'''}, \ldots, u_n >, < v_j, \ldots, v_m >)$, in which $i \leq i''' \leq n$ and $1 \leq j \leq m$, may contain frequent pairs. Hence, these pairs have to be collected and the frequent ones are added to the set of 2-grams.

All the g-frequent pairs are given as input to the discovery phase, as described by Algorithm 1. This means that one location in text may contribute to several pairs in the set, as we can see in the following example. If we assume that the frequency threshold is 2 and the pairs "I saw" and "he saw" occur in the text, the following pairs are g-frequent.

```
<nom, pron> <saw, see, past, v>
<nom, pron> <see, past, v>
<nom, pron> <past, v>
<nom, pron> <v>
<nom> <saw, see, past, v>
<nom> <see, past, v>
```

```
<nom> <past, v>
<nom> <v>
<pron> <saw, see, past, v>
<pron> <see, past, v>
<pron> <past, v>
<pron> <v>
```

All the subsets are necessary, since in longer sequences more general feature vectors may be needed to make a sequence frequent. Each subset is indexed by a unique integer, which is passed as input to the discovery phase. Hence the discovery phase cannot use the knowledge that some subset is a generalization of some others. After the discovery phase (as in Algorithm 1), the resulting sequences are not maximal g-sequences, but for each g-frequent sequence also all the more general variations of the sequence are returned. For instance, a sequence

```
<nom, pron> <saw, see, past, v> <det> <ball, nom, n>
```

also produces 23 other sequences, e.g.

```
<pron> <saw, see, past, v> <det> <nom, n>
<pron> <v> <det> <n>
```

As the maximal g-frequent sequences are a compact representation, the more general sequences should be pruned away from the result. Compared to the discovery phase, this pruning can be done fast and efficiently. Moreover, it is rather easy to find all the more general and shorter subsequences that occur more frequently. In our sample document collection (Fig. 1), the following maximal generalized frequent sequences can be found. All of them have a frequency 2, i.e., they occur in two documents.

```
<det> <green, green, abs, a> <ball, ball, nom, n>

<det> <red, red, abs, a> <ball, ball, nom, n> <abs, a>

<the, the, nil, det> <abs, a> <ball, ball, nom, n> <was, be, past, v> <abs,a>

<nom, pron> <saw, see, past, v> <det> <ball, nom, n>
```

4 Discussion

Both the basic method for finding maximal frequent word sequences and the extension for finding generalized sequences have been implemented in Perl. The basic method has been tested using, e.g., the Reuters-21578 news collection, and the method is designed — and further developed — to cope with large document collections. It is not clear, however, how far we can come with the extension. At the moment, it has been tested with a toy example only, and analysing any larger collection would probably need development of more efficient data structures.

The special characteristics of the linguistic features includes the large variation of the frequency of the features. It seems to be that in any text collection, however large, half of the words occur only once, while features like the part of speech form closed sets with a rather small number of distinct values. Hence, discovering patterns based on the frequency, treating all the features uniformly, may not be adequate.

Acknowledgements. This work is supported by the Academy of Finland (project 50959; DoReMi – Document Management, Information Retrieval, and Text Mining).

References

[1] Helena Ahonen. Knowledge discovery in documents by extracting frequent word sequences. *Library Trends*, 48(1):160–181, 1999. Special Issue on Knowledge Discovery in Databases.

[2] Helena Ahonen. Finding all maximal frequent sequences in text. In *ICML99 Workshop, Machine Learning in Text Data Analysis*, Bled, Slovenia, 1999.

[3] Helena Ahonen-Myka, Oskari Heinonen, Mika Klemettinen, and A. Inkeri Verkamo. Finding co-occurring text phrases by combining sequence and frequent set discovery. In Ronen Feldman, editor, *Proceedings of 16th International Joint Conference on Artificial Intelligence IJCAI-99 Workshop on Text Mining: Foundations, Techniques and Applications*, pages 1–9, Stockholm, Sweden, 1999.

[4] Rakesh Agrawal and Ramakrishnan Srikant. Mining sequential patterns. In *International Conference on Data Engineering*, March 1995.

[5] Heikki Mannila, Hannu Toivonen, and A. Inkeri Verkamo. Discovering frequent episodes in sequences. In *Proceedings of the First International Conference on Knowledge Discovery and Data Mining (KDD'95)*, pages 210–215, Montreal, Canada, August 1995.

[6] Rakesh Agrawal, Heikki Mannila, Ramakrishnan Srikant, Hannu Toivonen, and A. Inkeri Verkamo. Fast discovery of association rules. In Usama M. Fayyad, Gregory Piatetsky-Shapiro, Padhraic Smyth, and Ramasamy Uthurusamy, editors, *Advances in Knowledge Discovery and Data Mining*, pages 307–328. AAAI Press, Menlo Park, California, USA, 1996.

Pattern Detection and Discovery: The Case of Music Data Mining

Pierre-Yves Rolland[1,2] and Jean-Gabriel Ganascia[1]

[1] Laboratoire d'Informatique de Paris 6 (LIP6)
[2] Atelier de Modélisation et de Prévision, Université d'Aix-Marseille III,
15/19 allée Claude Forbin, 13627 Aix en Provence cedex 1, France
P_Y_Rolland@yahoo.com

Abstract. In this paper the problem of automatically detecting (or extracting, inducing, discovering) patterns from music data, is addressed. More specifically, approaches for extracting "sequential patterns" from sequences of notes (and rests) are presented and commented. Peculiarities of music data have direct impact on the very nature of pattern extraction and, correlatively, on approaches and algorithms for carrying it out. This impact is analyzed and paralleled with other kinds of data. Applications of musical pattern detection are covered, ranging from music analysis to music information retrieval.

1 Introduction

Music has been a problem (or application) area for data mining approaches and technologies for a number of years. It has however not been publicized, particularly, in data mining literature, as much as more traditional areas such as market basket analysis.

In musical data mining one seeks to "detect", "discover", "extract" or "induce[1]" of melodic or harmonic (sequential) patterns in given sets of composed or improvised works. There are many different applications, ranging from musical analysis to music generating systems [Rowe 1993; Ramalho et al. 1999] to the more recent applications in content-based music information retrieval (see final section). In music analysis, understanding and characterizing style is the core motivation of melodic pattern-based analysis, where the uncovering of recurring structures and of their organization in works offers a way to gain insights into the style or techniques of composers and improvisers. For instance, in a very large study [Owens 1974] musicologist T. Owens characterized the techniques and style of Charlie Parker through the nature and organization of recurrent melodic patterns ('formulae') in the jazz saxophone player's improvised playing. He extracted a hierarchically classified lexicon of 193 such formulae from a corpus of about 250 Parker solos (examples appear in 1). Many studies such as Owens' have been carried out on corpuses of both improvised music, e.g. Davis, Coltrane, and composed music, e.g. Debussy. These studies often take years for

---

[1] In the music problem area, pattern discovery is often called "pattern induction"

D.J. Hand et al. (Eds.): Pattern Detection and Discovery, LNAI 2447, pp. 190–198, 2002.
© Springer-Verlag Berlin Heidelberg 2002

a human analyist to carry out, as s/he must deal manually with the large combi-
natorics (see next section) inherent to the sequential pattern extraction process.
Hence, in melodic pattern-oriented music analysis, just like for other sequential
pattern extraction (SPE) contexts, there is a strong motivation in seeking to
automate the process.

Fig. 1. A few of Charlie Parker's patterns from the lexicon extracted by musicologist
T. Owens (1974). The hierarchical pattern identificators ('1Ac', '1Bb', etc.) are Owens'.

2   Semi-formal View

Although it is not the aim of this paper to provide details about algorithms
for music data mining, it is nevertheless important to give some clear basic
definitions about the data to be mined and the patterns to be extracted. (A
complete and formal set of definitions can be found in [Rolland 2001b]). It should
be noted that one of the things missing in most literature pertaining to music
data mining is precise definitions, even informal.

In seeking to discover patterns in music data, we take part in the more
general sub-area of data mining called sequential data mining, where data is
organized in sequences. In this case, there is a natural ordering (time in the case
of music) over data, and a database — called sequential database — is made of
one or several sequences. A sequence is an ordered collection of individual data
structures (possibly mere numbers or strings) called the sequence's *elements*. In
music a sequence is generally a musical work (composition, improvization) and
elements are notes, rests and possibly chords. Areas other than music where data
comes under the form of sequences include (a) other kinds of temporal data, for

instance in finance, telecommunications or multimedia (sound, speech, video, etc.); and (b) molecular biology and genomics/genetics.

Addressing the most usual situation in music data mining, we are concerned with the automated extraction of particular sequential patterns which roughly correspond to the intuitive notions of regularity and local recurrence in sequences. Although precise definitions will be given later on, we can for the moment give the following semi-formal definition. A *sequential pattern* is a set of segments from the sequential database which share a significant degree of resemblance: for pairs of segments in the pattern, the similarity between the two segments can be measured numerically and is above a given threshold (in that case the two segments are said to be *equipollent*). The particular similarity structure — i.e. which, among all possible pairs of segments, are required to be equipollent — depends on the classes of patterns sought. One may wish to extract patterns where all segment pairs are equipollent, but in the FlExPat approach [Rolland 2001b] that other similarity structures such as star-type patterns (see Figure 2) are more efficient yet yielding pertinent results.

Each segment is called an *occurrence* of the pattern. A pattern is characterized *extensionally* by a list of segments as was just introduced. It can also be characterized *intensionally* by a *prototype*: a structure that is representative of the set of segments it 'summarizes'. (We will focus here on the case where prototypes are sequential structures of the same form as the pattern's segments. In many application contexts, the prototype is in fact one of these segments, viz. a melodic passage from one of the music works). We can now semi-formally define SPE: *Given a sequential database over an alphabet Σ, a similarity function over Σ^* and an equipollence threshold, sequential pattern extraction consists of finding all sequential patterns that exist in the database.* Every extracted pattern is characterized extentionaly, intensionally, or both.

These definitions imply that, central to any SPE approach is a model of similarity between segments, which can be formalized as a *similarity function*. Such a function *simil*(Segments s,s') can numerically measure the similarity between any two segments from the sequential database, generally yielding a normalized value between 0 (maximum difference) and 1 (maximum similarity). An *equipollence threshold* T is chosen according to some strategy and segment pairs (s,s') we will be interested in are such that $simil(s,s') \geq T$.

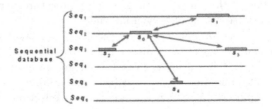

Fig. 2. An illustration of the notion of star-type pattern. Segment s0 (called the star's center) is equipollent to all other segments in the pattern. Other equipollence relations between other pairs of segments among s1, ..., s4 may or not exist.

We are dealing with patterns here rather than with *models*, according to the distinction made by Hand et al (2000), and Hand, Mannila, and Smyth (2001). Also, *pattern extraction* is concerned here rather than pattern detection (or discovery) in the broadest sense. We propose to make the folowing distinction between detection/discovery on one hand, and extraction on the other. Discovery can in some cases mean mere pattern detection — viz detection of the fact that there are local regularities in some part of the data. Extraction, on the other hand, implies that, in addition to their detection, the patterns are explicitly available at the end of the extraction process, described in some language.

3   Key Issues in Music Pattern Extraction

A number of papers have presented methods and techniques for automatically extracting sequential patterns from music sequences — mostly from monodic (i.e. monophonic melodic) sequences. In this section, in addition to presenting sequential pattern extraction approaches and algorithms, we analyze the impact that the peculiarities of music have on them. It will be interesting to study how some of the key issues also appear in other problem areas, and whether the solutions that have been found for music could be applied to these other areas and vice versa.

We will not mention here all existing SPE systems for music (A more complete presentation can be found e.g. in [Rolland & Ganascia 1999]). Rather, the following representative list of systems will be covered, not so much with the aim of describing them in details but rather for illustrating the key issues addressed in the next subsections.

EMI [Cope 1991], a system for automatically composing western tonal music using SPE.

The system by [Pennycook et al. 1993], a music-listening and -improvisation system that uses SPE to interact with other (human) performers in jazz style.

Cypher [Rowe 1993], a music-listening and -generating system similar to Pennycook et al.'s, but without a target musical genre such as jazz.

The system for experimenting with different representations in music analysis designed by Smaill et al. (1993)

Imprology (which implements the FlExPat algorithm) [Rolland 2001b]: an object-oriented environment for experimenting with music representation, pattern extraction and pattern matching in music analysis.

In Figure 3 we schematize the various steps, explicit or implicit, that compose any of the mentioned music SPE algorithms.

4   Data and Knowledge Representation

In most situations there are several possible ways to represent sequences and their elements. Additionally, many application domains are knowledge-intensive, meaning that artificial systems must possess vast amounts of domain-specific knowledge to achieve satisfactory performance. Music is among these domains

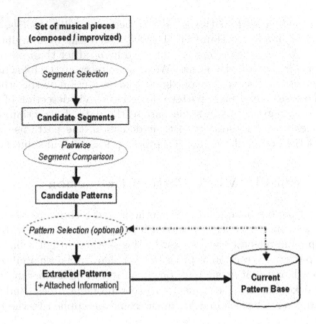

Fig. 3. Unified processing scheme for music SPE systems.

[Widmer 1994]. It has been seen by several researchers (including the author) to imply that representing music for SPE (typically: melodic sequences of notes and rests) should not be limited to usual basic descriptions such as absolute pitch (e.g. "D# 4") and relative duration (e.g. "eight-note"). Figure 4 shows some musical descriptions that have been found to intervene in music perception, recognition, rememberance and cognition in general. (Sources are music theory and music psychology works at large).

Fig. 4. Some musical descriptions, ranging from more abstract ones (bottom) to more concrete ones (top).

Among the various existing SPE systems, there is a very large spectrum as regards music representation. Some use very basic, trivial ones, which has of course a number of practical advantages; however it has been widely recognized that, for any task (like music SPE) that involves complex cognitive representations when it is carried out by a human, if a system is to carry out the task it should be able to compute and manipulate as much of these representations.

In Imprology, based on music theory and music psychology works at large, and on empirical experimentation, we have proposed a flexible framework for representing musical sequences based on multiple simultaneous descriptions. These descriptions are organized according to their level of abstraction, as illustrated in Figure 4. We have designed and implemented algorithms for incrementally computing all chosen descriptions based on the initial, basic representation of notes and rests. (It can be remarked that part of the concepts in this general framework apply directly to domains other than music.) Within the SPE process, a user-controlled phase is inserted where the initial representation is enriched (with additional descriptions)—or changed. This enrichment or change is based (generically) on psychological works on music perception and cognition, and music theoretical works as mentioned above. It is on the final representation obtained that pattern extraction is carried out (see Figure 5). We have automated within Imprology this representation enrichment/change phase for melodic sequences, which typically have an initial MIDI-type representation. From these we identified and classified a set of eligible user-selectable musical descriptions at various levels of abstraction, which can be derived from the basic (initial) descriptions of melodies and notes/rests. Additional descriptions are always derived from initial descriptions, or from any combination of initial and already derived descriptions. For instance, the "BackwardChromaticInterval" description, which is the number of semitones between a note and its predecessor, is computed from the "AbsolutePitch" description of the note under consideration and of its predecessor.

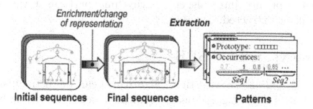

Fig. 5. Inserting a representation enrichment/change stage before the actual SPE process in the Imprology system.

5   Similarity Measurement

The computational schemes for measuring the similarity between candidate pairs of segments (similarity function as introduced above) have a major influence on

SPE algorithms. In fact, as will be seen there is a trade-off between computational efficiency and the richness of the similarity notion captured by the system —
which is intrinsically related to the computational schemes for similarity mea-
surement. Among the simplest of these schemes are boolean ones: the segments
compared are determined either identical ('True') or different ('False'). With
such a notion of similarity, some music SPE systems have been designed (Smaill
et al. 1993) using very simple (brute-force) pattern extraction algorithms. However it has been widely recognized that such systems, which find *exactly* repeating
segments, are completely inappropriate in many musical contexts. In fact, in music, repetitions are hardly ever exact, because of *variation* and *ornamentation*
(a bit like mutation and evolution in molecular biology).

At the other end of the spectrum, very rich models of sequential similarity
have been designed for music contexts based on generalizations of the edit distance [Levenshtein 1965] framework. The important idea is that, to measure
the similarity between two segments, one should find the best *correspondence
structure* between them, i.e. between their elements. Cypher uses the basic Levenshtein distance (with fixed, unit costs). Pennycook et al.'s system adds non-
standard edit operations (*fragmentations* and *consolidations*); costs functions
are variable but only use two simple descriptions for notes (chromatic interval
and relative duration ratio), with equal weights. In Imprology we have further
generalized and flexibilized this model in designing the *multi-description valued
edit model* (MVEM). It has been designed to be able to deal with representations of sequences of notes and rests using multiple simultaneous descriptions.
In MVEM edit operators are: *replacement, insertion, deletion, fragmentation,
multideletion,* and *generalized substitution*. MVEM uses variable cost functions
that take the form of weighted sums of terms, each relating to a single description in the (multiple) representation. The user or calling program can adjust the
weight attributed to each description, allowing them in particular to take different viewpoints over the data. For instance, on musical data, by giving greater
weights to temporal descriptions — durations, metrics, etc.— than to frequential
descriptions — pitches, intervals, etc.—, rhythmic patterns, rather than melodic
patterns, will be extracted.

6   Online vs. Offline SPE

In music SPE, *online* (i.e. real time) algorithms can be distinguished from *off-
line* algorithms. Offline algorithms are typically used in music analysis contexts,
where the set of music material under study is fully accessible when SPE is
performed (e.g. [Cambouropoulos 1998; Cope 1991; Rolland 1999]). Online algorithms are used in real-time performance contexts, where the material on which
pattern processing is carried out 'unveils' progressively in time (e.g. [Penny-
cook et al. 1993; Rowe 1993]. For instance, some interactive accompaniment and
improvisation systems attempt incrementally to extract, in real time, melodic
patterns from their fellow performers' playing. The systems then reuse the extracted patterns, possibly after adaptation, in their own playing. A system's
ability to 'capture' another (human) performer's musical gestures can make a

very strong impression on human listeners, such as ability is generaly seen as a complex, human-like one in cognitive terms.

Segmentation is an issue in any music SPE context. It consists of 'cutting' data sequences to define candidate segments whose similarity with other segments will be computed. But segmentation is particularly difficult and critical in online SPE, because it has to be carried out in real time. In [Pennycook et al. 1993; Rowe 1993], simple strategies are used for real-time segmentation of a real-time melodic stream.

However, it is widely recognized that there are often several different, all musically pertinent, ways to segment a given musical sequence. Additionally, one musically pertinent segmentation for one person can be non-pertinent for another. For these reasons, segmentation — even non real-time — in SPE contexts remains an open issue.

7   Discussion and Perspectives

Several SPE systems have shown satisfactory results. For instance, pieces have been generated by EMI that interestingly imitate real composers' style. Also, a number of the formulae of Owens' lexicon have been 'rediscovered' by FlExPat from the same data. Additionaly, FlExPat has discovered new formulae (that had not been listed by Owens) and these formulae have been validated as pertinent by human experts.

Research projects are still carried out for designing new SPE algorithms for general music applications such as the ones mentioned above. Nevertheless, what we see as the key direction in current music data mining research is related to music information retrieval (MIR). Concepts, techniques and systems are being developed for retrieving music information from collections of music content, using some query-based or navigation-based paradigm. Among the most promising approaches is the WYHIWYG paradigm (*What You Hum Is What You Get*) [Rolland et al. 1999], also refered to as Query-by-Humming. content-based music retrieval as in Desirable data to include in such a user model include. MIR is a very fast-growing research area (see e.g. the International Symposium on Music Information Retrieval which will be held in its 3rd edition in 2002) as there are diverse contexts where MIR can be used, some with great commercial stake. These range from searches for pieces in music libraries to consumer-oriented e-commerce of music.

Among key topics in current MIR research is looking for ways to keep search time (query processing) under a few seconds although the databases searched get larger and larger. Intelligent indexing seems to be a potential solution to this problem: It can be observed that, in WYHIWYG queries, user often sing/hum/whistle song themes rather than particular verses. Themes, in turn, are by definition passages that are repeated two or more times, possibly with some variation, throughout a given music work. Thus SPE, as a database preprocessing stage, would allow to automatically extract themes from music pieces. Carrying out the search algorithm (pattern matching) on such a theme database would dramatically decrease search time. Of course, in case the user doesn't find what s/he was looking for in this 'fast-search' mode, s/he can always re-run a 'traditional' full-database search using the same query.

References

Cambouropoulos, E. 1998. Musical Parallelism and Melodic Segmentation. In Proceedings of the XII Colloquium of Musical Informatics, Gorizia, Italy.

Cope, D. 1991. Computers and Musical Style. Oxford: Oxford University Press.

Hand D.J., Blunt G., Kelly M.G.,and Adams N.M. 2000. Data mining for fun and profit. Statistical Science, 15, 111-131.

Hand D.J., Mannila H., and Smyth P. 2001. Principles of data mining, MIT Press.

Levenshtein, V.I. 1965. Binary codes capable of correcting deletions, insertions and reversals. Cybernetics and Control Theory 10(8): 707-710.

Owens, T. 1974. Charlie Parker: Techniques of Improvisation. Ph.D. Thesis, Dept. of Music, UCLA.

Pennycook, B., D.R. Stammen, and D. Reynolds. 1993. Toward a computer model of a jazz improviser. In Proceedings of the 1993 International Computer Music Conference, pp. 228-231. San Francisco: International Computer Music Association.

Ramalho, G., Rolland, P.Y., and Ganascia, J.G., 1999. An Artificially Intelligent Jazz Performer. Journal of New Music Research 28:2.

Rolland, P.Y. 2001a. Introduction to Pattern Processing in Music Analysis and Creation. *Computers and the Humanities* Vol. 35, No. 1. Kluwer Academic Publishers.

Rolland, P.Y. 2001b. FlExPat: Flexible Extraction of Sequential Patterns. Proceedings IEEE International Conference on Data Mining (IEEE ICDM'01). San Jose - Silicon Valley, California, USA, November 29 -December 2, 2001.

Rolland, P.Y. 2001c. Adaptive User Modeling in a Content-Based Music Retrieval System. Proceedings 2nd International Symposium on Music Information Retrieval (ISMIR'01). Bloomington, Illinois, USA, October 15-17, 2001.

Rolland, P.Y., Cambouropoulos, E., Wiggins, G. (editors). 2001. Pattern Processing in Music Analysis and Creation. Special Issue of Journal: Computers and the Humanities Vol. 35, No. 1. Kluwer Academic Publishers

Rolland, P.Y., Ganascia, J.G. 1999. Musical Pattern Extraction and Similarity Assessment. In Miranda, E. (ed.). Readings in Music and Artificial Intelligence. New York and London: Gordon & Breach - Harwood Academic Publishers.

Rolland, P.Y., Raskinis, G., Ganascia, J.G. 1999. Musical Content-Based Retrieval: an Overview of the Melodiscov Approach and System. In Proceedings of the Seventh ACM International Multimedia Conference, Orlando, November 1999. pp. 81-84

Rowe, R. 1993. Interactive Music Systems. Cambridge: MIT Press.

Smaill, A., Wiggins, G. & Harris, M. (1993) Hierarchical Music Representation for Composition and Analysis. (1993). In Computational Musicology, edited by Bel, B. Vecchione, B. Special Issue of Computer and the Humanities 27(1). North Holland Pub. Co.

Widmer, G. 1994. The Synergy of Music Theory and AI: Learning Multi-Level Expressive Interpretation. Extended version of paper appearing in Proc. AAAI'94 (11th NCAI)

Discovery of Core Episodes from Sequences*
Using Generalization for Defragmentation of Rule Sets

Frank Höppner

Department of Computer Science
University of Applied Sciences Wolfenbüttel
Salzdahlumer Str. 46/48
D-38302 Wolfenbüttel, Germany
frank.hoeppner@ieee.org

Abstract. We consider the problem of knowledge induction from sequential or temporal data. Patterns and rules in such data can be detected using methods adopted from association rule mining. The resulting set of rules is usually too large to be inspected manually. We show that (amongst other reasons) the inadequacy of the pattern space is often responsible for many of these patterns: If the true relationship in the data is fragmented by the pattern space, it cannot show up as a peak of high pattern density, but the data is divided among many different patterns, often difficult to distinguish from incidental patterns. To overcome this fragmentation, we identify core patterns that are shared among specialized patterns. The core patterns are then generalized by selecting a subset of specialized patterns and combining them disjunctively. The generalized patterns can be used to reduce the size of the set of patterns. We show some experiments for the case of labeled interval sequences, where patterns consist of a set of labeled intervals and their temporal relationships expressed via Allen's interval logic.

1 Introduction

Although we will concentrate on (a special kind of) sequential data later, the problem that will be discussed in this paper is well-known wherever association rule mining techniques are applied. Association rule discovery [1] yields large sets of rules "if A then B with probability p", where A and B are patterns from an application-specific pattern space. For instance, in market basket analysis, the pattern space consists of all sets of items that can be purchased in a supermarket. Association rules are considered as potentially useful for knowledge discovery purposes since rules are easily understood by humans.

One problem with all these techniques (regardless of whether they are applied to itemsets [1], event sequences [12], calendar patterns [11], or interval sequences [6]) is the size of the rule set, which is often too big to be scanned and evaluated manually. Usually, an expert of the field has to think over every rule carefully to

---

* This work has been supported by the Deutsche Forschungsgemeinschaft (DFG) under grant no. Kl 648/1.

D.J. Hand et al. (Eds.): Pattern Detection and Discovery, LNAI 2447, pp. 199–213, 2002.
© Springer-Verlag Berlin Heidelberg 2002

decide whether the discovered correlation is incidental, well-known, or indicates something potentially new. Providing too many rules to the expert overburdens him or her quickly – and thus limits the usefulness of rule mining for knowledge discovery.

Why are there so many rules? The rules are generated from a set of patterns (in some pattern space P) that occur more often than a certain threshold, which is why they are called *frequent* patterns. (The number of occurrences of a pattern is denoted as the *support* of a pattern.) Given a subpattern relationship $\sqsubseteq$ for patterns in P, for any two frequent patterns A and B with $A \sqsubseteq B$ a rule $A \to B$ can be generated. It is interpeted as "whenever we observe pattern A we will also observe pattern B with probability p". For instance, if the pattern space consists of sets of ingredients of recipes, from a frequent pattern $B = \{water, flour, sugar, salt, eggs\}$ any subset A can be chosen to derive a rule $A \to B$, e.g. $\{water, flour\} \to \{water, flour, eggs\}$ (often abbreviated as "water, flour $\to$ eggs"). Obviously, the number of rules can be much larger than the large set of frequent patterns. Besides those patterns that are found due to true dependencies in the data (those we want to discover), incidental co-occurrences (that appear more often than the minimum support threshold) also introduce many frequent patterns (and rules).

One can find many approaches in the literature to reduce the number of rules and patterns, e.g. by concentrating on maximal patterns [2], by restricting to closed patterns [13], or by incorporating additional user information [10].

Other extensions of association rule mining have also been proposed. For the case of market basket data, product hierarchies or taxonomies have been suggested to generalize items "orange juice", "apple juice", "cherry juice" simply to "juice" in order to obtain stronger rules. It may be appropriate to distinguish the different kinds of juices for some associations, while it may be better to *generalize* juices in others. These problems have been solved by introducing a priori knowledge about the product taxonomy [15,5]. However, we cannot be sure that our taxonomy contains all useful generalizations: Do we want to consider "tomato juice" in our *is-a* hierarchy as a "juice" or introduce another level to distinguish between vegetables and fruit juices?

What is the relation of this to the problem of large rule sets? The product hierarchies seem to address a completely different phenomenon at first glance. On second thought, the stronger *generalized juice* rules make a number of *specialized orange/apple/cherry juice* rules obsolete. In absence of the juice generalization we have a number of only moderately strong rules and miss the strong "true relationship" in the data due to an inadequacy or incompleteness of the pattern space. It would be possible to learn the generalized terms automatically, if our pattern space would allow not only simple rules

$$ADE \to FG, \quad CDE \to FG, \quad BDE \to FH, \quad \dots \tag{1}$$

but also disjunctive combinations like

$$(A \vee B \vee C)DE \to F(G \vee H) \tag{2}$$

If a taxonomy is given, it may happen that $(A \lor B \lor C)$ matches a generalized term in our taxonomy (but this is not guaranteed, of course). If the *true association* in the data is given by (2) or negative patterns "A but no B", and our pattern space does not contain such patterns we will observe *fragments* of the true pattern only – as in (1). Thus, we call a pattern space inadequate when we want to express the fact that there may be relationships in the data that are not representable by hypotheses from the pattern space.

One may want to conclude that the choice of the pattern space should be thought over, however, the pattern space has usually been chosen after carefully balancing its modelling capabilities and the computational cost of searching it efficiently. We do not want to put the pattern space in question, because the consideration of all possible disjunctive combinations would increase the cost of rule mining dramatically. We consider the question of whether it is possible to identify the true patterns by generalization of the fragmented patterns. Specialization and generalization are considered to be two of the most basic techniques used by the brain to generate new rules. While the association rule mining process can be seen as a *specialization step* (every possible rule is enumerated), a second phase of *generalizing* the specialized patterns seems to be worthwhile.

The outline of the paper is as follows. Section 2 will motivate our interest in labeled interval sequences. To judge whether a rule's specialization or generalization should be considered we use the J-measure, which is discussed in Sect. 3. In case of sequential data it makes sense to concentrate on meaningful subsequences in the data (Sect. 4) rather than rules alone. The value of a subsequence will be given by the value of the rule that can be extracted from the subsequence (Sect. 5). We think that this approach is advantageous for the expert, since the number of such sequences will be much smaller and more easy to verify than a set of rules. We concentrate on an interesting subset of meaningful subsequences (Sect. 6), which we will use as a basis for generalization (Sect. 7). An example will be given in Sect. 8.

2   Motivation for Interval Sequences

We are interested in sequential data and more precisely in labeled interval sequences. Since such sequences are not widely used, we want to motivate them briefly and introduce the pattern space we are considering.

We think of labeled interval sequences as a natural generalization of discrete (static) variables to time-varying domains. If time is not considered explicitly, an attribute v of an object is simply denoted by a single value x. The value $v = x$ holds at the time of recording, but not necessarily in the past or future. Whenever the value of v changes, the time interval of observing $v = x$ is ended and a new interval $v = y$ starts. If the value of v is unknown at time t, then we have no interval I in the sequel with $t \in I$. The development of the variable over time (as much as we know about it) is thus characterized by a sequence of labeled intervals, where the labels denote the variables value x and the interval denotes a time period I where $v(t) = x$ holds for $t \in I$.

Such interval sequences can be given naturally (medical patient data, insurance contracts, etc.), but may also be obtained from abstracting some other raw data. For instance, in case of event sequences it may be useful to aggregate similar events to intervals of equal event density. Although in the example we have considered discrete variables only (for continuous variables it is very likely that the interval width becomes zero), this representation is also extremely useful to capture high-level descriptions of continuous-valued variables like time series.

In fact, our main concern is the discovery of dependencies in multivariate time series. Although a complex system is difficult to forecast or model as a whole, such systems (or subsystems) cycle very often through a number of internal states that lead to repetitions of certain patterns in the observed variables. Observing or discovering these patterns may help a human to resolve the underlying causal relationships (if there are any). Rather than trying to explain the behaviour of the variables *globally*, we therefore seek for *local dependencies* or *local patterns* that can be observed frequently[1]. Having found such dependencies, an expert in the field may examine what has been found and judge its importance or relevance. Since correlations do not necessarily indicate cause-effect relationships, the final judgement by an expert is very important. On the other hand, an expert in the field is not necessarily familiar with whatever kind of analysis we are going to use, therefore it is important to obtain results that are easily understandable for the domain expert. Therefore we use artificial (discrete) variables whose values address qualitative aspects of the slope or curvature of the signal. The big advantage of this conversion to labeled interval sequences is the fact that this representation corresponds pretty well to the perception of a time series profile by a human. Humans argue in terms of shape or visual appearance, which are attributes that are less easily distorted by noise.

A pattern in our pattern space consist of a set of labeled intervals and their temporal relationships $\mathcal{I}$ expressed via Allen's interval logic (cf. figure 1). Patterns may then describe, for instance, subsequences in multivariate time series such as "interval *air-pressure convex* overlaps interval *air-pressure increases*, both intervals are contained in an interval *wind-strength increases*". The pattern captures only qualitative aspects of the curves as well as qualitative interval relationships, they are therefore well-suited to compensate dilation and translation effects which occur very often in practice (and are difficult to handle in many other approaches). Formally, given a set of n intervals $[b_i, f_i]$ with labels $s_i \in \mathcal{S}$, a temporal pattern of size n is defined by a pair (s, R), where $s : \{1, .., n\} \rightarrow \mathcal{S}$ maps index i to the corresponding interval label, and $R \in \mathcal{I}^{n \times n}$ denotes the relationship between $[b_i, f_i]$ and $[b_j, f_j]$. We use the notation $|P|$ to denote the size of a pattern P (number of intervals). For pattern frequency estimation, we choose a sliding window $\mathcal{W}$ of a certain width that is slid along the sequence. The support of a pattern is given by integrating the time period in which the pattern is visible. For more details on the pattern space, we refer the interested

---

[1] The term *frequently* should not be taken too seriously, we refer to patterns that occur in a certain percentage p of all observations, but p can be as small as 1%, for example.

reader to [9,8]. For the abstraction of the raw time series to interval sequences
we use scale-space filtering and wavelet techniques [7].

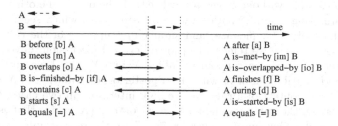

Fig. 1. Allen's interval relationships $\mathcal{I}$ (Abbreviations in parenthesis).

While the pattern space has very useful properties for our purposes, it still
may be inadequate for observing certain relationships in the data: Consider the
following pattern "signal B starts to increase after any signal but A has started
to increase". There is no equivalent element in our pattern space that corresponds to this relationship because some qualitative statements about interval
end-points are missing nor can we express a negated term ("any but A"). So we
observe only a number of artefacts (like "C-increase overlaps B-increase" or "C-
increase before B-increase" or "D-increase meets B-increase" etc.), which are
by no means incidental patterns (although they could easily be misclassified as
such): They are specializations of a pattern that is not contained in the pattern
space – and since the support of the true pattern is shared among the artefacts
they cannot show up that clearly as the true pattern would.

3   The Rule Evaluation Measure

Association rule mining techniques enumerate all rules that fulfil a certain minimum support and confidence. In a naive approach to generalization we could
do the same, that is, enumerate all generalizations that fulfil these conditions. A
temporal pattern of dimension k has k different labels and $k \cdot (k-1)/2$ interval
relationships (the remaining interval relationships can be determined uniquely).
Potentially, we would have to generalize any of these $k + k \cdot (k-1)/2$ values for
each pattern. Such a "bottom-up" approach to generalization increases the computational burden significantly and increases the number of rules even further.

Instead, we are interested in reducing the number of rules, that is, apply
generalization to the resulting rule set in order to reduce its size and at the same
time improve its overall value. There are two important properties of a rule,
its specificity or applicability (denoted by the support of the premise pattern)
and its goodness-of-fit or confidence. Generalization of a rule makes it less specific (which is good, because it can be applied more often) but may reduce the
goodness-of-fit at the same time (which is bad, because the rule holds in fewer

cases) – so, has the rule improved overall or not? This is the difficult question a rule evaluation measure has to decide. We want the measure to decide whether the rule itself or its specialization or generalization should be kept in the set of rules – preferably without a bias towards either the one or the other.

We are using the information-theoretic J-measure, which seems to balance the goodness-of-fit (confidence) and simplicity (support) of the rule very well, due to lack of space we refer to [14] for a detailed discussion. Given a rule "if $Y = y$ then $X = x$" on random variables X and Y, the J-measure compares the a priori distribution of X with the a posteriori distribution of X given that $Y = y$. In the context of a rule, we are only interested in two cases, given that $Y = y$, either the rule was right ($X = x$) or not ($X = \bar{x}$), that is, we only consider the distribution of X over $\{x, \bar{x}\}$. Then, the relative information

$$j(X|Y = y) = \sum_{z \in \{x, \bar{x}\}} Pr(X = z|Y = y) \log_2 \left(\frac{Pr(X = z|Y = y)}{Pr(X = z)} \right)$$

yields the *instantaneous* information that $Y = y$ provides about X (j is also known as the Kullbach-Leibler distance or cross-entropy). When applying the rule multiple times, on average we have the information $J(X|Y = y) := Pr(Y = y) \cdot j(X|Y = y)$. The value of J is bounded by ≈ 0.53 bit [14]. In the context of temporal patterns in labeled interval sequences we have already obtained promising results using this measure when specializing the rules with quantitative constraints [8].

We understand rules $P \to R$ such that P and R are patterns with $P \sqsubseteq R$ (cf. Sect. 1). As a consequence, the probability of observing the rule pattern R without observing the premise pattern P is zero, which may appear a bit unusual when compared to static rules. However, with temporal patterns we want the rule to resolve the temporal relationships between intervals in premise and conclusion. Therefore, the conclusion pattern must contain the premise pattern in order to specify the temporal relationship between premise and conclusion intervals. In a rule $P \to R$ we thus speak of a premise pattern P and a *rule pattern* R rather than a conclusion pattern. The J-value of a rule tells us how informative the premise is w.r.t. an occurrence of the rule pattern.

For the J-value of such rules the following holds: Given a premise P of a rule $P \to R$. Then among the rules with the highest J-value there is a rule pattern of size $|P| + 1$. Given a rule $P \to R$, among the rules $P \to R'$ with $P \sqsubseteq R' \sqsubseteq R$ and lowest J-values there is $R' = R$. (Proofs straightforward but omitted due to space limitations.)

4   Frequent Episodes

We have illustrated in Sect. 2 that we assume repetitions of certain patterns in the observed time series. One could therefore argue that the discovery of such repeating patterns (or subsequences) is our main concern, rather than enumerating rules. Indeed, from a single frequent pattern (or sequence) the number of generated rules increases exponentially with the length of the pattern, thus, concentrating on the patterns seems to be much less effort. Moreover, the interestingness of a rule depends on the variables it uses: The best-rated rule is not necessarily most helpful for the expert since it may use unknown variables in the premise or predict well-known variables. By providing the sequences, an expert can decide on his own which intervals he wants to have in the premise and which ones in the conclusion before considering the derived rules.

Although we have fewer patterns than rules, it is still true that for every k-pattern we have $2^k - 1$ subpatterns, thus, the number of subpatterns is still large. Some authors consider only patterns that are maximal, that is, there is no other frequent pattern that contains it as a subpattern (e.g. [2]). While this reduces the number of patterns significantly, we obtain for every incidental occurrence of an interval X in the vicinity of a maximal pattern a new maximal pattern. Such incidental occurrences do not add any information to the maximal frequent patterns, therefore it seems to be more promising to concentrate on the *maximal patterns among the interesting patterns* (without having yet defined interestingness for sequences).

Which sequences are meaningful? It is safe to assume that not every frequent pattern corresponds to a meaningful repetition of some internal states in a system. Following an idea by Cohen and Adams [3], a meaningful sequence, called episode in the following, is characterized by the fact that the sequence gets more and more deterministic: At the beginning of an episode, we are not quite sure what we will observe next, but the more elements of the episode we have seen, the more we are certain about what kind of episode we are currently observing and it becomes more easy to predict the next observation. At the end of an episode we are again uncertain how to continue. In [3] entropy is used to measure the goodness-of-fit as the sequence develops, thus the end of an episode is recognized by a increase in the entropy (cf. algorithm in Fig. 2).

This appealing idea seems to be helpful in distinguishing incidental patterns from meaningful episodes. However, as we have discussed in Sect. 3, a uniform a priori belief is not very well suited to compare goodness-of-fit values for different sequences. Cohen and Adams perform some normalization to cope with this. We simply apply the goodness-of-fit term j of the J-measure for this purpose.

By considering episodes rather than sequences we are more robust against incidental occurrences of intervals. Since we do not want to consider such incidental patterns during generalizations of rules, we use the set of episodes rather than the set of frequent sequences for further processing.

5   Evaluating Episodes

We are still lacking an episode evaluation measure. From their definition we know that the goodness-of-fit increases as the episode becomes longer. We want to rate episodes by their suitability to create strong rules out of them, and thus goodness-of-fit alone is not suited to measure the usefulness of an episode, the average applicability of a rule is also important (as we have discussed in Sect. 3).

We want to use the J-values for rules that can be obtained from an episode to rank the episode itself. How many rules can be derived from a k-episode R? We will use the alternative notation $P \to_R C$ for a rule $P \to R$ with conclusion pattern C under rule pattern R for notational convenience. Let us divide R into a premise P_i and conclusion part C_i, such that $|P_i| = i$ and $|C_i| = |R| - i$. There are $|R| - 1$ possibilities for this subdivision. Any pair of subsequences (besides the empty sequence) $P' \sqsubseteq P$ and $C' \sqsubseteq C$ will do for a potential rule $P' \to C'$. We have $|P|$ intervals in the premise and $|C|$ intervals in the conclusion, and thus for each subdivision $(2^{|P|} - 1) \cdot (2^{|C|} - 1)$ rules. Which rules shall we use to rank the episode then?

If we simply use the maximum of all J-values of all rules

$$J_R = \max_{i \in \{1,..,|R|-1\}} \max_{C' \sqsubseteq C_i} \max_{P' \sqsubseteq P_i} J_{P' \to_R C'} \qquad (3)$$

it impossible to distinguish between different developments of the J-value with an increasing length of the sequence. Two sequences R and R' get the same ranking even if the maximum J-value is obtained for different points of subdivision i. But smaller values of i are preferable (given the same J-value), because then a shorter prefix is necessary to reliably predict the continuation of the sequence. Therefore, rather than using a single number we use a tuple of $|R| - 1$ J-values to rank an episode, such as

$$J_R^+ = \left(\max_{C' \sqsubseteq C_i} \max_{P' \sqsubseteq P_i} J_{P' \to_R C'} \right)_{i \in \{1,..,|R|-1\}} \qquad (4)$$

Now, $J_R^+[i]$ denotes the J-value of the best subrule of $P_i \to R$. (Note that $J_R^+[i]$ is not necessarily increasing with i as the goodness-of-fit does.)

We are not quite satisfied with this, because a strong relationship between a short i-prefix of R and the 1-suffix of R would yield consistently high J-values for all $J_R[i']$ with $i' > i$, because for $i' > i$ there is always a subrule that contains the i-prefix in the premise and the 1-suffix in the conclusion. Therefore, our final choice for the episode evaluation is an $(|R| - 1)$-tuple such that $J_R[i]$ yields the minimal J-value that can be obtained from a subpattern of P_i for any conclusion $C' \sqsubseteq C_i$:

$$J_R^- = \left(\min_{C' \sqsubseteq C_i} \max_{P' \sqsubseteq P_i} J_{P' \to_R C'} \right)_{i \in \{1,..,|R|-1\}} \qquad (5)$$

A ranking of episodes can be obtained by sorting the $J_R^-[i]$ values in decreasing order[2] and comparing the tuples (of varying size for varying length of the episode) lexicographically. Note that $[J_R^-[i], J_R^+[i]]$ provides an interval of J-values in which any rule (with any conclusion $C' \sqsubseteq C_i$) and optimized premise $R' \sqsubseteq R_i$ will fall.

---

[2] J_R^- is monotonically increasing with i, thus sorting corresponds to reversing the order.

Fortunately, the calculation of the $[J_R^-, J_R^+]$ intervals is log-linear in the number of frequent episodes and thus can be done efficiently. For every k-pattern P we store a $(k-1)$-vector of J-values $P_{minJ}[\cdot]$ and $P_{maxJ}[\cdot]$. We sort all frequent patterns such that the prefix property is preserved (if P is a prefix of Q then P comes before Q, see e.g. [9]). Then we identify episodes from patterns as described in the previous section by running once through the patterns in this order. At any time we keep all i-prefixes $Q[i]$ of a k-pattern R and initialize $R_{minJ}[i] = R_{maxJ}[i] = J(R|Q[i])$. The J-vector thus contains the J-values of rule patterns R that are subdivided into premise and conclusion at position i (cf. Fig. 2).

```
1  let F be a sorted list of frequent patterns of size 1 ≤ k ≤ K        O(|F| log |F|)
2  for P ∈ F let P_maxJ[·] and P_minJ[·] be a (|P| − 1)-tuple
3  let Q be an empty vector of patterns
4  fetch first pattern R ∈ F
5  do                                                                    O(|F|)
6      k = |R|; Q[k] = R;
7      if k ≤ 2 ∨ j(Q[k − 1]|Q[k − 2]) < j(R|Q[k − 1])
8      then
9              append R to E
10             for i = 1 to k − 1 do R_maxJ[i] = R_minJ[i] = J(R|Q[i]) od
11             fetch next pattern R ∈ F
12     else
13             fetch next pattern R ∈ F until |R| ≤ k
14     fi
15 until all patterns R ∈ F are processed
16 now E is a sorted list of frequent episodes
```

Fig. 2. Determining episodes among frequent patterns. For every episode R, $R_{minJ}[i]$ (and $R_{maxJ}[i]$) contains the J-value of the rule that is obtained by subdividing the rule pattern R into premise and conclusion at position i.

For $k = 2$ the values $R_{minJ}[1]$ and $R_{maxJ}[1]$ already correspond to $J_R^-[i]$ and $J_R^+[i]$, because only a single rule can be derived from a 2-episode. Then, we iterate over all episodes R of length 3 and generate all 2-subepisodes P obtained by removing one of the 3 intervals. The R_{minJ} and R_{maxJ} vectors contain the J-values for rules that use all intervals in R, from P_{minJ} and P_{maxJ} we obtain the J-values for rules that use only 2 out of 3 possible intervals. The J-value of the best rule is obtained by taking the respective maximum of J-values. Now, the J-vectors contain the J-value of the best subrule of R and we continue to process episodes of size 4 and so forth (cf. Fig. 3). For the R_{minJ} values we make use of the fact that the rule with $C' = C_i$ has the smallest J-value (cf. Sect. 3).

```
 1  let ε be a sorted list of frequent episodes of size 1 ≤ k ≤ K
 2  for k = 3 to K do
 3      for R ∈ ε ∧ |R| = k do                              both for-loops together: O(ε)
 4          for i = 1 to k − 1 do
 5              let Q ⊑ P where the iᵗʰ interval of P has been removed
 6              if Q ∈ ε                                     search in ordered set: O(log ε)
 7              then
 8                      for j = 1 to k − 1 do
 9                          if j ≤ k then jj = k − 1 else jj = k fi
10                          R_maxJ[j] = max(R_maxJ[j], Q_maxJ[jj])
11                          if i ≤ j then R_minJ[j] = max(R_minJ[j], R_minJ[jj]) fi
12                      od
13              fi
14          od
15      od
16  od
```

Fig. 3. Evaluating the $J_R^-[i]$ and $J_R^+[i]$ vectors for episodes.

6   Core Episodes

The J-measure will serve us in ranking the episodes by the average information content of the rules that we may derive from them. But we are not only interested in a sequential ranking of all episodes, but also in reducing the number of episodes that we have to consider. The idea of maximal episodes is that any subepisode can be generated from a superepisode, therefore it makes no sense to enumerate all subepisodes. We have already indicated in Sect. 4, maximal episodes represent large sets of episodes, but do not lead to the most frequent interesting episodes since they usually contain noise. This is true if the minimum support threshold is the only property available for episodes. Now, we consider all rules that can be generated from an episode. Given an episode, from any superepisode we can generate all those rules that can be generated from the episode itself. Thus, when considering the maximum J-value that can be obtained according to (3) or (4), a superepisode cannot have smaller J-values than any of its subepisodes – which is basically the same situation as before where we had no episode measure. (Noisy patterns are "preferred" due to the fact that they have more subpatterns.) This is the reason for not using the maximum of all rules but the definition (5).

We consider episodes as distinguished if the full episode is needed to obtain the best J-value; that is, there is a subdivision point i and no subrule of $P_i \rightarrow_R C_i$ yields a better J-value. If $P_i \rightarrow_R C_i$ contains incidental intervals we can improve the value of the rule by removing them. But if the best J-value is obtained by using all intervals, it is very likely that all of these intervals are meaningful in this context. Therefore we call such an episode a *core episode*. It provides the core for many maximal superepisodes but the J-value cannot be improved by them. Therefore we believe that the core episodes are close to those patterns that are caused by the repetitive cycling through internal states of a system. The core property can be determined by the algorithm in Fig. 3 if we additionally store a pointer to the pattern that provides $J_{minJ}[i]$. If for some pattern P and some i this pointer leads us to P itself, we have identified a core episode.

With maximal episodes no maximal superepisodes exist (by definition), but core episodes may have superepisodes that are core episodes. This is due to the fact that a long episode may obtain its core property from a subdivision point that is larger than any subdivision point of the subrule. An episode ABC (interval relationships omitted) may be a core episode from $i = 2$, that is, the rule $AB \to C$ has a better J-value than any of its subpatterns. However, if ABC is a prefix of an even longer episode $ABCDE$, the superepisode may also be a core episode for $i = 3$. Among the set of core episodes we therefore consider only the maximal core episodes (that is, core episodes that have no core superepisode).

7 Episode Generalization

At this point we have identified a set of maximal core episodes, which is much smaller than the set of episodes or even maximal episodes. By definition, every interval in a core episode seems to be meaningful, since it cannot be removed without decreasing the J-value. So, is it sufficient to present only the maximal core episodes to the domain expert?

Our answer is yes, given that the pattern space is powerful enough to express all relationships in the data. As we have mentioned in the introduction, we think that this pattern space adequacy cannot be guaranteed – except in rare cases. Assuming pattern space inadequacy, the answer is no. We have already discussed several examples: If the relationship is decomposed into many different fragments it is very likely that none of the fragments itself yields a strong rule and thus none of these fragments becomes a maximal core episodes. Nevertheless, such episodes carry valuable information despite their low J-value – if they are considered in the context of similar episodes.

Our approach to solve this problem is the generalization of episodes to new (disjunctive) longer episodes. For a naive approach (cf. Sect. 3) the computation effort is far too big to be feasible. But the pattern space is usually chosen carefully, taking the kind of expected patterns into account. So what we obtain is by no means a random fragmentation of the true pattern, but it is very likely that certain aspects of the true relationship can be captured by an episode in the pattern space. From a computational perspective it is much more promising to start generalization from such a near-miss episode than generalizing everything. Our hypothesis is that maximal core episodes provide such near-miss patterns.

Thus, we will use the core episodes as the starting point for generalization. As before, the J-measure plays an important role in judging the usefulness (or acceptance) of a possible generalization. It will only be accepted if the J-value of the episode can be improved. Thus, a generalization itself becomes a (generalized) maximal core episode. As a positive side effect, the existence of a new (longer) core episode prunes other core episodes which are now subepisodes of the generalization. Thus, generalization helps to reduce the number of maximal core episodes as well as improving the rating of the remaining core episodes.

We always select the best-so-far maximal core episode to be generalized next. Once the episode has been generalized, it will not be considered again for generalization. Let R be such an episode. Generalization of R will be done incrementally, starting from episodes (not only core episodes) of length $|R|+1$, then $|R|+2$, etc. Thus we start with collecting in S all episodes of length $|R|+1$ that contain R as a subepisode. Since we want to improve the J-rating of episode R, we start from the largest possible subdivision point i down to 1 (cf. episode ranking in Sect. 5). For each value of i we try to find a subset $G \subseteq S$ such that the disjunctive combination of episodes in G maximizes the J-value[3]. If the maximal J-value is higher than that of the core episodes $J_R[i]$, generalization was succesful and G is considered as a new maximal core episode.

Finding a subset G of S requires to estimate the support of the disjunctive combination of episodes in G, which usually requires another database pass. We are not allowed to simply add the support values of the single episodes since some sliding window positions may contribute to multiple of the episodes in G. However, during frequent pattern enumeration we maintain a condensed representation of the pattern locations, namely a list of intervals in which the patterns are visible [6,9]. This representation can be used to unite and intersect support sets efficiently and thus to find G more efficiently.

Having found some i and G such that the J-value is improved over the core episode, we mark the episodes in G as being part of a generalized core pattern and mark all subpatterns of them, too. This is to exclude maximal core patterns that are subepisodes of the generalized core pattern from further generalization.

8   Experiment

We want to consider an artificial test data set we have used earlier in [8]. We prefer to show some results using an artificially generated data set over a real data set, because only in this case do we *know* about the true patterns in the data. The following description of the data set is taken from [8].

> We have generated a test data set where we have randomly toggled three states A, B, and C at discrete time points in $\{1, 2, ..., 9000\}$ with probability 0.2, yielding a sequence with 2838 states[4]. Whenever we have encountered a situation where only A is active during the sequence generation, we generate with probability 0.3 a 4-pattern A

---

[3] The number of superepisodes was very moderate (< 20), so we used complete search in our experiments, which is not feasible for large sets of superepisodes. Our current objective is to develop a pruning technique for the search of the best core generalization.

[4] That is, at each point in time we introduce a new start or end point (depending on whether the state was currently active or not) with probability 0.2. This is done for each of the variables independently.

meets B, B before C, and C overlaps a second B instance. The length
and gaps in the pattern were chosen randomly out of $\{1, 2, 3\}$. [...]
We consider the artificially embedded pattern and any subpattern
consisting of at least 3 states as interesting.

Thus, the 3 rules $A \rightarrow BCD$, $AB \rightarrow CD$ and $ABC \rightarrow D$ were considered as
interesting in [8]. In terms of episodes, this corresponds to a single interesting
episode $ABCD$ (interval relationship omitted). For a window width of 16 and a
very low minimum support threshold of 0.1% we obtain 17484 frequent patterns
with up to 10 intervals. From the low support value, the comparatively small
database, and the fact that we have only a single true relationship in the data,
we expect many incidental patterns that are not meaningful. Considering only
meaningful episodes rather than all patterns reduces the set to 8579 (49%),
among them are 2425 maximal episodes (13.9%). However, we have found only
669 core episodes (3.8%), among them 515 maximal core episodes (2.9%). As
expected, the best maximal core episode is our artifically embedded pattern:

$$A \rightarrow_{[0.0004, 0.01]} B \rightarrow_{[0.36, 0.46]} C \rightarrow_{[0.47, 0.47]} B$$

The intervals denote the range of possible J-values for rules extracted from the
episode. For instance, if the expert agrees with the episode and wants to use
it as a rule, whatever conclusion he may select from the two last symbols, by
selecting an appropriate subset from the premise he will obtain a J-value within
$[0.36, 0.46]$. The gap between the two rightmost J-intervals is smaller than the
the gap between the two leftmost intervals, from which one can conclude that
the premise "A" alone does not provide useful rules but in combination with B.

As already mentioned, the data set from [8] was not designed to illustrate
generalization. Therefore we were surprised to find a meaningful generalization
of the $ABCD$ sequence. The following 11 superepisodes (depicted graphically)
were considered during generalization:

$$\boxed{X}\ \boxed{A\,|\,B}\ \boxed{C} \qquad \text{for } X \in \{A, B, C\}, \qquad (6)$$
$$\boxed{B}$$

$$\boxed{X\,|\,A\,|\,B}\ \boxed{C} \qquad \text{for } X \in \{B, C\}, \qquad (7)$$
$$\boxed{B}$$

$$\boxed{A\,|\,B\,|\,A}\ \boxed{C} \qquad\qquad (8)$$
$$\boxed{B}$$

$$\boxed{A\,|\,B}\ \boxed{C} \qquad \text{for } X \in \{B, C\}, \qquad (9)$$
$$\boxed{X}\qquad\ \boxed{B}$$

$$\boxed{A\ \ |\,B}\ \boxed{C} \qquad \text{for } X \in \{B, C\}, \qquad (10)$$
$$\boxed{X}\qquad\quad\ \boxed{B}$$

$$\boxed{A\ |\,B}\ \boxed{C} \qquad\qquad (11)$$
$$\boxed{C}\qquad\quad\ \boxed{B}$$

While a disjunctive combination of these specialized rules did not increase
the J-value of the rule $ABC \rightarrow B$ it did for $AB \rightarrow CB$. The increase in the

minimum J-value was only moderately (increase from 0.3618 to 0.3675), but nevertheless interesting to interpret. The generalization used 6 out of the possible 11 superepisodes. None of the episodes in (6) was used: For any meaningful pattern it is very likely to observe any of the labels A, B, or C in a *before* relationship if the window is sufficiently large. Consequently, these labels are not useful for predicting the continuation of the sequence and thus do not improve the J-value of a rule – it makes sense to discard these episodes during generalization. Also episode (8) was not considered: From the description how the patterns were generated, the second A instance tells us that the first 3 intervals were not generated according to the explanation given above – the pattern is thus incidental and correctly discarded. All other episodes, besides (9) for $X = B$ were used for generalization. While we have no explanation why the case $X = B$ is excluded, the used episodes (7), (9), (10) and (11) share a common property: There is a B or C instance with non-empty intersection with the first A instance, but the right bound of the A instance is never included in this intersection. This summary comes pretty close to the original formulation

> Whenever we have encountered a situation where only A is active during the sequence generation, we generate [...]

With the used pattern space we are not able to express that no other intervals besides A are observable at some point in time. However, if we observe that a B and/or C instance has ended just before an A instance is ended, it seems to be much more likely that the condition "only A is active" holds at the end of A. This observation increases the goodness-of-fit of the specialized rules but cannot be reflected by a single episode. Only by generalization we can find this relationship and overcome the inadequacy of the pattern space.

9   Conclusion

In this paper we have examined how to shrink the amount of "discovered knowledge" (patterns or rules obtained from association rule mining) that has to be inspected by an expert of the field manually. There are three reasons for an unnecessary high number of rules: (a) subrules are enumerated, (b) incidental occurrences produce further variants of rules, (c) inadequacy of pattern space. While the first two have been addressed before in the literature, the pattern space inadequacy has not been addressed explicitly. By pattern space inadequacy we refer to the fact that the definition of the search space does not necessarily contain all relationships in the data. We assume that this inadequacy holds more often than not – even for the artificial data set used in Sect. 8, which has been designed *before* we thought about pattern space inadequacy, this phenomenon can be observed. We have proposed a way to overcome these problems by restricting ourselves to core episodes rather than all frequent patterns (addressing (b)), considering only the maximal core episodes for episode enumeration (addressing (a)), and providing a scheme for core episode generalization (to address (c)). The preliminary results we have achieved so far are promising.

References

[1] R. Agrawal, H. Mannila, R. Srikant, H. Toivonen, and A. I. Verkamo. Fast discovery of association rules. In [4], chapter 12, pages 307–328. MIT Press, 1996.

[2] H. Ahonen-Myka. Finding all maximal frequent sequences in text. In D. Mladenic and M. Grobelnik, editors, *Proc. of the ICML99 Workshop on Machine Learning in Text Data Analysis*, pages 11–17, 1999.

[3] P. R. Cohen and N. Adams. An algorithm for segmenting categorical time series into meaningful episodes. In *Proc. of the 4th Int. Symp. on Intelligent Data Analysis*, number 2189 in LNAI, pages 197–205. Springer, 2001.

[4] U. M. Fayyad, G. Piatetsky-Shapiro, P. Smyth, and R. Uthurusamy, editors. *Advances in Knowledge Discovery and Data Mining*. MIT Press, 1996.

[5] J. Han and Y. Fu. Discovery of multiple-level association rules from large databases. In *Proc. of the 21st Int. Conf. on Very Large Databases*, pages 420–431, 1995.

[6] F. Höppner. Discovery of temporal patterns – learning rules about the qualitative behaviour of time series. In *Proc. of the 5th Europ. Conf. on Principles of Data Mining and Knowl. Discovery*, number 2168 in LNAI, pages 192–203, Freiburg, Germany, Sept. 2001. Springer.

[7] F. Höppner. Learning dependencies in multivariate time series. In *Proc. of the ECAI'02 Workshop on Knowledge Discovery from (Spatio-) Temporal Data*, pages 25–31, Lyon, France, July 2002.

[8] F. Höppner and F. Klawonn. Finding informative rules in interval sequences. In *Proc. of the 4th Int. Symp. on Intelligent Data Analysis*, volume 2189 of *LNCS*, pages 123–132, Lissabon, Portugal, Sept. 2001. Springer.

[9] F. Höppner and F. Klawonn. Learning rules about the development of variables over time. In C. T. Leondes, editor, *Intelligent Systems: Technology and Applications*, volume IV, chapter 9, pages 201–228. CRC Press, 2003. To appear.

[10] M. Klemettinen, H. Mannila, P. Ronkainen, H. Toivonen, and A. I. Verkamo. Finding interesting rules from large sets of discovered association rules. In *Proc. of the 3rd Int. Conf. on Inform. and Knowl. Management*, pages 401–407, 1994.

[11] Y. Li, S. Wang, and S. Jajodia. Discovering temporal patterns in multiple granularities. In J. Roddick and K. Hornsby, editors, *Proc. of the 1st Int. Workshop on Temporal, Spatial, and Spatio-Temporal Data Mining*, number 2007 in LNAI, pages 5–19, Lyon, France, Sept. 2000. Springer.

[12] H. Mannila, H. Toivonen, and A. I. Verkamo. Discovering frequent episodes in sequences. In *Proc. of the 1st Int. Conf. on Knowl. Discovery and Data Mining*, pages 210–215, Menlo Park, Calif., 1995.

[13] N. Pasquier, Y. Bastide, R. Taouil, and L. Lakhal. Discovering frequent closed itemsets for association rules. In *Proc. of Int. Conf. on Database Theory*, number 1540 in LNCS, pages 398–416. Springer, 1999.

[14] P. Smyth and R. M. Goodman. An information theoretic approach to rule induction from databases. *IEEE Trans. on Knowledge and Data Engineering*, 4(4):301–316, Aug. 1992.

[15] R. Srikant and R. Agrawal. Mining generalized association rules. In *Proc. of the 21st Int. Conf. on Very Large Databases*, pages 407–419, 1995.

Patterns of Dependencies in Dynamic Multivariate Data

Ursula Gather[1], Roland Fried[1], Michael Imhoff[2], and Claudia Becker[1]

[1] Department of Statistics, University of Dortmund, Vogelpothsweg 87,
44221 Dortmund, Germany
{gather, fried, cbecker}@statistik.uni-dortmund.de
http://www.statistik.uni-dortmund.de/lehrst/msind/msind_e.htm
[2] Surgical Department, Community Hospital Dortmund, Beurhausstr. 40,
44137 Dortmund, Germany
mike@imhoff.de

Abstract. In intensive care, clinical information systems permanently record more than one hundred time dependent variables. Besides the aim of recognising patterns like outliers, level changes and trends in such high-dimensional time series, it is important to reduce their dimension and to understand the possibly time-varying dependencies between the variables. We discuss statistical procedures which are able to detect patterns of dependencies within multivariate time series.

1 Introduction

Modern technical possibilities allow for simultaneous recording of many variables at high sampling frequencies. Possibly there are interactions between the observed variables at various time lags and we have to treat the data as multivariate time series. Often the data contain strong dynamic structures, that are unknown in advance. In intensive care for instance, physiological variables, laboratory data, device parameters etc. are observed for each critically ill patient. The appropriate analysis and online monitoring of this enormous amount of dynamic data is essential for suitable bedside decision support in time critical situations [1]. Thus, methods for automatic abstraction of the dynamical information into clinical relevant patterns are needed. Physicians typically select some of the observed variables and base their decisions on patterns like level shifts and trends detected in them. Statistics offers alternatives such as dynamic factor analysis providing a few latent variables which describe the dynamical information in the data appropriately. However, to get reliable and interpretable results by any method for dimension reduction without substantial loss of information we need to understand the relations between the variables.

We analyse 11-variate time series describing the hemodynamic system as measured in intensive care, consisting of arterial and pulmonary artery blood pressures (diastolic, systolic, mean), denoted by APD, APS, APM, PAPD, PAPS, PAPM, central venous pressure (CVP), heart rate (HR), pulse (PULS),

D.J. Hand et al. (Eds.): Pattern Detection and Discovery, LNAI 2447, pp. 214–226, 2002.
© Springer-Verlag Berlin Heidelberg 2002

blood temperature (TEMP), and pulsoximetry (SPO2). We first discuss graphical models, that reveal the partial correlation structure within multivariate time series [2], [3]. Performing such graphical analyses for different physiological states allows to characterise distinct health states by distinct correlation structures [4]. Next, we briefly explain how the information obtained from graphical models can be used to enhance dynamic factor modelling. Finally, methods for detecting non-linear dependence-patterns (SIR; [5]) are transferred into the context of high-dimensional medical time series.

2   Graphical Models

2.1   Partial Correlation Graphs

Graphical models are almost standard nowadays for investigating relations within multivariate data [6], [7], [8], [9]. A *graph* $G = (V, E)$ consists of a finite set of *vertices* V and a set of *edges* $E \subset V \times V$. A visualization can be accomplished by drawing a circle for each vertex and connecting each pair a, b of vertices for which $(a, b) \in E$ or $(b, a) \in E$. If only one of these pairs is included in E, e.g. (a, b), then a directed edge (*arrow*) is drawn from a to b. If both pairs are included in E an undirected edge (*line*) is used. An arrow specifies a directed influence, while a line stands for a symmetrical relation.

When analysing the relations between the vital signs of an individual we should consider the time series structure of the measurements, which are not independent. Partial correlation graphs for multivariate time series introduced by Brillinger [2] and Dahlhaus [3] visualize the (partial) *linear relations* between the components of a multivariate time series. Such an analysis of symmetric relations is a first step to get a better understanding of the underlying dependence structure. We just note that directed graphical models providing information on possible causalities have also been developed recently [10].

In the following, let $X(t) = (X_1(t), \ldots, X_k(t))'$, $t \in \mathbb{Z}$, be a multivariate stationary time series of dimension k. Suppose that the autocovariance function

$$\gamma_{ab}(h) = Cov(X_a(t+h), X_b(t)), \ h \in \mathbb{Z}, \tag{1}$$

is absolutely summable with respect to all time lags h for all $a, b \in \{1, \ldots, k\}$. Then the *cross-spectrum* between the time series $X_a = \{X_a(t), t \in \mathbb{Z}\}$ and $X_b = \{X_b(t), t \in \mathbb{Z}\}$ is defined as the Fourier-transform of their covariance function $\gamma_{ab}(h), h \in \mathbb{Z}$,

$$f_{ab}(\lambda) = f_{X_a X_b}(\lambda) = \frac{1}{2\pi} \sum_{h=-\infty}^{\infty} \gamma_{ab}(h) \exp(-i\lambda h). \tag{2}$$

This defines a decomposition of the covariance function γ_{ab} into periodic functions of frequencies λ. The variables X_a and X_b are uncorrelated at all time lags h iff $f_{ab}(\lambda)$ equals zero for all frequencies [11].

In order to distinguish between direct and induced linear relations between two component series X_a and X_b, the linear effects of the remaining components $Y = \{(X_j(t), j \neq a, b), t \in \mathbb{Z}\}$ have to be controlled. For this the optimal $\mu_a \in \mathbb{R}$ and the optimal filter $d_a(h), h \in \mathbb{Z}$, have to be determined, such that the quadratic distance

$$E\left[X_a(t) - \mu_a(t) - \sum_h d_a(h)Y(t-h)\right]^2 \tag{3}$$

is minimized. Let $\epsilon_a = \{\epsilon_a(t), t \in \mathbb{Z}\}$ be the residuals obtained from this, and calculate ϵ_b from X_b in the same way. The correlations between these residual series at all time lags define the partial correlation structure between X_a and X_b after eliminating the linear effects of the other variables. The *partial cross-spectrum* $f_{X_a X_b \cdot Y}(\lambda)$ between X_a and X_b is then defined as the cross-spectrum between ϵ_a and ϵ_b, while the *partial spectral coherence* is a standardization hereof,

$$R_{ab \cdot Y}(\lambda) = \frac{f_{ab \cdot Y}(\lambda)}{\sqrt{f_{aa \cdot Y}(\lambda) f_{bb \cdot Y}(\lambda)}}. \tag{4}$$

In a *partial correlation graph* for a multivariate time series the vertices $a = 1, \ldots, k$ are the components of the time series and an undirected edge between two vertices a and b is omitted, $(a, b) \notin E$, whenever the correlations between the residual series ϵ_a and ϵ_b are zero for all time lags. This is equivalent to $f_{X_a X_b \cdot Y}(\lambda) = 0$ for all frequencies $\lambda \in \mathbb{R}$. This defining property is called the *pairwise Markov property*. Partial correlation graphs illustrate the indirect character of some marginal correlations as two variables a and b are marginally related (possibly via other variables) if they are *connected* by a *path*, i.e. if vertices $a = a_0, a_1, \ldots, a_k = b$, $k \geq 0$, exist, such that there is an edge between each subsequent pair of vertices.

If the spectral density matrix is regular at all frequencies the pairwise Markov property implies the *global Markov property* for this kind of graphical model, that is a stronger property in general [3]. The latter property says that two sets of variables $A \subset V$ and $B \subset V$ have zero partial correlations given the linear effects of a set of variables $C \subset V$ if C separates A and B in G, i.e. if any path between two variables $a \in A$ and $b \in B$ necessarily contains at least one variable $c \in C$. In other words, the variables in A and B are not related if the effects of the separating subset C are controlled. This allows to illustrate zero partial correlations by means of separation properties of a partial correlation graph. The subset C may contain less than all the remaining variables which allows to identify important variables more clearly.

2.2   Application to Physiologic Time Series

In the following we apply partial correlation graphs for multivariate time series to physiologic variables representing the hemodynamic system. Online-monitoring data was acquired in one minute intervals from 25 consecutive critically ill patients with pulmonary artery catheters for extended hemodynamic monitoring, amounting to 129943 sets of observations altogether, i.e. about 5200 observation times are available for each patient on the average. We compare "empirical associations" found by statistical analysis to "physiological associations" based on medical knowledge. Physiological association means that a change in one physiological variable leads to a change in another physiological variable. This term does not imply any causal, linear or non-linear relation. Not even a direction or an ordering in time is expected for it as e.g. during volume depletion an increase of CVP and PAPM will typically lead to an increase in APM and a decrease in HR, while in congestive heart failure under high doses of inotropes an increase in HR may typically lead to an increase in APM.

The program "Spectrum" developed by Dahlhaus and Eichler [12] estimates the cross-spectra by a nonparametric kernel estimator. The partial spectral coherences are estimated from these cross-spectra using an inversion formula derived by Dahlhaus [3]. Then a decision has to be made on whether the partial spectral coherences may be identical to zero because sampling variability always causes estimates to be distinct from zero. Spectrum also constructs an approximate bound for the 95%-percentile of the maximal squared estimated partial spectral coherence under the assumption that the true partial spectral coherence equals zero. This allows one to perform an approximate 5%-test for the hypothesis of partial uncorrelatedness of two variables by comparing the estimated partial spectral coherence with this bound. However, it is well-known that different relations between physiological variables may have distinct strengths. The strength of a relation can be understood as an expected relative change in one of the variables when the other one changes by a certain relative amount. Therefore, we classify the relations as high, medium, low and zero partial correlation on the basis of the area under the estimated partial spectral coherence $R_{X_a X_b \cdot Y}$. This area can be measured by the partial mutual information between the time series X_a and X_b, which is defined by

$$- \frac{1}{2\pi} \int \log\{1 - |R_{ab \cdot Y}(\lambda)|^2\} d\lambda, \tag{5}$$

(see Granger and Hatanaka [13] and Brillinger [2]) or by variants of this. Then we construct partial correlation graphs using gradually distinct edges to represent the strength of the empirical relation.

Figure 1 displays the resulting partial correlation graph for the hemodynamic system derived from one patient. Neglecting the many low relations we can identify some groups of strongly related variables from the graph. High partial correlations exist between the systolic, diastolic and mean arterial pressure, between the heart rate and the pulse, as well as between the systolic, diastolic and mean pulmonary artery pressure. The central venous pressure is mainly related to the pulmonary artery pressures. The blood temperature and the pulsoximetry seem to be rather isolated. These findings are similar for all patients analysed in our case-study, particularly the subgroups of highly partially correlated variables could always be identified. Some differences were found w.r.t.

the relations of the other variables (CVP, Temp, SPO2) when performing such a one-step model selection, but the partial correlation graphs obtained by a step-wise search strategy [14] closely resembled the one shown here in most of the cases. For some of the patients medium to high partial correlations were found between temperature and pulsoximetry, which were not expected by medical knowledge, but turned out to be artifacts as the measuring instrument used for pulsoximetry is influenced by the temperature.

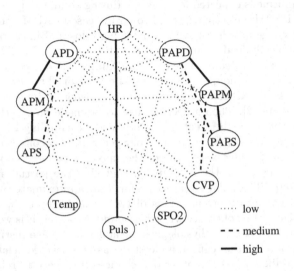

Fig. 1. Partial correlation graph for the hemodynamic system of one patient.

Such a partitioning of the variables into strongly related subgroups can be used to reduce the number of variables which have to be considered for online monitoring, i.e. for variable selection. From Figure 1 we see that APM provides a lot of information (measured via the partial mutual information) on both APD and APS, while both APD and APS provide less information on each other. Selecting APM from this strongly related subgroup and neglecting APD and APS for clinical monitoring is therefore meaningful from a statistical point of view. In the same way we see that choosing one variable out of PAPD, PAPM and PAPS should be sufficient. Indeed, physicians often select the mean pressures APM and PAPM as well as the heart rate. Thus, they choose exactly that variable out of each subgroup identified by the partial correlation model which has the strongest relation to the other variables. This is due to the nature of the mean pressures, which are "in between" the diastolic and systolic pressures. Hence, the statistical results agree with medical knowledge.

2.3   Distinguishing Clinical States

A patient can suffer from several clinical states like pulmonary hypertension, septic shock, congestive heart failure and vasopressor support during his visit at an intensive care unit. Distinct clinical states are accompanied by different pathophysiological responses of the circulatory system. These changes may be supposed to result in differences in the interactions between the vital signs, particularly in the interactions between the groups of closely related variables identified above. Next we investigate, whether partial correlation graphs can detect differences in the status of the circulatory system. We drop systolic and diastolic pressures in the following calculations, which is in line with medical reasoning as systolic, diastolic and mean blood pressures are different representations of the same physiological process, i.e. a pulsatile blood pressure. The reason for dropping some of the variables is that in the previous data analysis the linear influences of *all* other variables have been subtracted. Elimination of the influences of APD and APS from APM e.g. may hide some relations of APM to other variables as the remaining variability is very low.

For each patient the predominant clinical state is determined for every time point as scored from the medical record using a set of rules used at the Hospital Dortmund. In the following we estimate the relations within a set of 'important variables' consisting of HR, SPO2, APM, PAPM, CVP and Temp for each state and for each patient separately.

Figure 2 summarizes the results of our data analysis using partial correlation graphs. Although the number of samples is small, there are obvious differences between the partial correlation patterns for distinct clinical states. Most of these differences can be explained by known physiological mechanisms. While for most states strong partial correlations could be found between heart rate and blood pressures as well as between the blood pressures, the status of *pulmonary hypertension* is predominantly associated with strong partial correlations between PAPM and CVP. This corresponds to the fact that this state can be characterized by an elevated PAPM. Since CVP also influences the right ventricle, one expects strong interactions between CVP and PAPM. On the other hand, the higher resistance within the pulmonary bloodstream attenuates the interactions between PAPM and APM as changes in PAPM will have a less than normal effect on left ventricular preload. In the status of *congestive heart failure* there are strong partial correlations between APM and PAPM. This can be explained by a failure of the left ventricle, that causes a decrease of APM as the heart is not able to push the blood forward (forward failure). In consequence there is a build-up in front of the left ventricle and thus an increase of PAPM (backward failure). For the clinical status of *vasopressure support* there are strong partial correlations between APM and PAPM, too. However, there are also medium to high partial correlations between HR and APM. This is due to the therapy the patient gets in this status, which affects the heart as well as the bloodstream and puts off the usual autonomous regulation of the circulatory system.

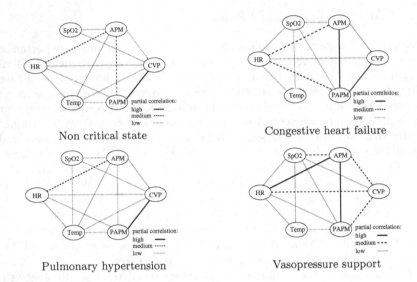

Fig. 2. Classification of the partial correlation patterns for distinct clinical states.

3   Factor Analysis

The detection of patterns like outliers, trends and level shifts in multivariate data becomes increasingly complex with increasing number of dimensions. Therefore, physicians usually base their decision on a subset of the variables which they regard as particularly important. Similarly, factor analytic methods allow to compress the data into a few latent variables which capture the essential information in the observed data as good as possible in a certain (statistical) sense. In this way a reduced number of factor series is obtained, that can be more easily overlooked by the physician than the observed series, the number of model parameters is reduced, and clinical relevant patterns are easier to detect also by an automatic data analysis. More precisely, in factor analysis it is assumed that there are a few, say l, latent variables called factors which drive the series and cause the correlations between the observable variables. For dynamic factor analysis of a multivariate time series $\{X(t), t \in \mathbb{Z}\}$ Peña and Box [15] propose the model

$$X(t) = \Lambda Z(t) + \varepsilon(t), \quad t \in \mathbb{Z}, \tag{6}$$

where Λ is a $k \times l$-matrix of loadings, $Z(t)$ are l-dimensional vectors of latent factors following a VARMA(p,q)-model, and $\{\varepsilon(t), t \in \mathbb{Z}\}$ is a k-dimensional process of Gaussian white noise with zero mean and arbitrary covariance matrix Σ_ϵ, which is independent of $\{Z(t), t \in \mathbb{Z}\}$. To get identifiability of the parameters, $\Lambda'\Lambda$ can be restricted to be the identity. If the factor series are independent, i.e. $Z(t)$ follows a VARMA(p,q)-model where all coefficient matrices are diagonal, then the time-lagged autocovariance matrices $\Gamma_X(h)$ of $\{X(t), t \in \mathbb{Z}\}$ are symmetrical for $h \geq 1$ and the columns of Λ can be chosen as the common eigenvectors of $\Gamma_X(h)$ while the eigenvalues $\gamma_i(h)$, $i = 1, \ldots, l$, are the diagonal elements of the autocovariance matrices $\Gamma_Z(h)$ of $\{Z(t), t \in \mathbb{Z}\}$ then. These findings can be used to identify a factor model for a given time series [15].

For the construction of an intelligent alarm system based on a dynamical factor analysis of the vital signs it is important that the factors are interpretable for the physician and that they reveal the important patterns found in the observable variables. Then we can simply monitor the factors and a summary measure of the model errors like the sum of the squared residuals for each time point.

One possibility to achieve better interpretability of the factors is to apply a rotation in the l-dimensional space. The previous findings obtained using graphical models justify separate factor analyses for the groups of strongly related variables, cf. Figure 1. Treating the subsets consisting of the arterial pressures, of the pulmonary artery pressures including the central venous pressure, and of the heart rate and the pulse simplifies the task of factor extraction from the hemodynamic system. We treat the blood temperature and the pulsoximetry separately as these do not show strong relations to the other variables. Since the variables of each group are measured on the same scale, we use the sample covariance matrices and calculate the eigenvalues and the eigenvectors for each group for the covariance matrices up to the time lag 4. We find one factor to be sufficient for each group. The resulting factor loadings are shown in Table 1. Very similar loadings are obtained if we analyze the variables jointly and use VARIMAX rotation to get better interpretable results. Hence, the factor loadings calculated in a factor analysis of all variables "identify" each of the rotated factors to belong to one of the subgroups. Thus, the results of both analyses nearly coincide. Analysing the variables in groups as is done here, however, affords less observations to get stable eigenvectors and provides better interpretable results as the loadings of the other variables not included in the respective group are exactly identical to zero.

Table 1. Factor loadings calculated for grouped variables.

Variable	factor 1	factor 2	factor 3
PAPD	0.3671	0	0
PAPM	0.5414	0	0
PAPS	0.6849	0	0
CVP	0.3211	0	0
APD	0	0.2449	0
APM	0	0.4776	0
APS	0	0.8438	0
HR	0	0	0.6964
PULS	0	0	0.7177

A closer inspection reveals that the calculated factors represent structural changes found in the corresponding component series better than any single variable [16]. Thus, dynamic factor modelling may facilitate the detection of patterns like level shifts and trends in multivariate physiological time series. Moreover, one may speculate that clinical relevant patterns of dependencies may be detected from an analysis of the partial correlations between the factor series similarly as in Section 2.3. For this aim, methods for estimating time-varying multivariate spectra [17] have to be further developed and adapted to the online monitoring context. Alternatively, moving window techniques can be applied to detect changes in the (partial) correlations between the factors [18].

4   Sliced Inverse Regression

4.1   A Dynamic Version of Sliced Inverse Regression

Restricting to linear relations as in the previous sections is not always appropriate. For this reason, Becker et al. [19] transfer sliced inverse regression (SIR) into the time series context to investigate possibly non-linear dependencies within multivariate time series. In its original form, SIR is a tool for dimension reduction in non-dynamic regression problems [5]. To explain the basic idea assume that $X = (X_1, \ldots, X_d)'$ is a set of explanatory variables (predictors), Y is a dependent variable, and ε an additional error variable. Instead of relating the whole set of predictors to Y via an unknown link function, i.e. taking $Y = g(X, \varepsilon)$, one assumes that it is sufficient to consider a lower-dimensional space, the so-called central dimension reduction (dr) subspace [20] $\mathcal{B}$ of dimension $r < d$, such that there is a function $f : \mathbb{R}^{r+1} \mapsto \mathbb{R}$ with

$$Y = f(\beta_1' X, \ldots, \beta_r' X, \varepsilon), \tag{7}$$

$$\text{where } \mathcal{B} = \text{span}[\beta_1, \ldots, \beta_r] \tag{8}$$

is a space of smallest possible dimension r such that (7) is satisfied. Hence, a reduction of the regressor space from d to r dimensions is supposed to be possible. SIR then estimates $\mathcal{B}$ using information contained in an estimate of the inverse regression curve $E(X|Y)$. The basic theorem underlying SIR states that under certain regularity assumptions the appropriately standardised inverse regression curve almost surely falls into a linear transform of $\mathcal{B}$. Reversely, the space in which $E(X|Y)$ is mainly spread out yields information on $\mathcal{B}$. To identify this space, the inverse regression curve $E(X|Y)$ is roughly approximated by a simple step function, and the space itself is estimated by means of a principal component analysis of the steps. Note that estimating the function f itself is not part of SIR but has to be performed afterwards. Hence, the SIR procedure is an "intermediate" in between projection pursuit regression (yielding an estimate of f together with a reduced space of projections) and unsupervised principal component analysis of the regressor X (yielding a reduced space without considering any information on f).

Becker et al. [19] suggest the following modification of the original SIR procedure to incorporate time series structure: Various lagged measurements of the variables are bound together to construct the regressor space, forming higher dimensional observations. Then the original SIR method is applied to this higher dimensional regressor. Formally, let $(Y(t), X(t)')'$ denote the observation of $(Y, X')'$ at time t. Then we can search for a dr subspace of the regressor space using a modified version of equation (7):

$$Y(t) = f(\beta_1' \widetilde{X}(t), \ldots, \beta_r' \widetilde{X}(t), \varepsilon(t)), \ t \in \mathbb{Z}, \tag{9}$$

where $\widetilde{X}(t) = (X(t)', Y(t-1), X(t-1)', \ldots, Y(t-p), X(t-p)')'$ if we want to take p time lags into account. The order p has to be chosen using preliminary information or by successive fitting of increasing orders. Experience shows that for online monitoring data observed in intensive care $p = 2$ is sufficient [21]. Effectively, by applying this dynamic version of SIR we assume a nonlinear transfer function model

$$Y(t) = g(X(t), \ldots, X(t-p), Y(t-1), \ldots, Y(t-p), \varepsilon(t)) \tag{10}$$

where the innovations $\varepsilon(t)$ form a white noise process. Here, it is usually supposed that feedback is not present in the system, meaning that Y does not influence the explanatory variables X and that the processes $\{\varepsilon(t), t \in \mathbb{Z}\}$ and $\{X(t), t \in \mathbb{Z}\}$ are independent. Appropriate dimension reduction in this model class is crucial to get an impression of g as finding a suitable transfer function and estimating its parameters is difficult even in case of a single explanatory variable [22]. For more details on non-linear models see [23].

4.2   Application to Physiologic Time Series

Dynamic SIR as described above is a powerful exploratory tool in data situations, where we expect that there is some dependence structure between the variables without knowing it in detail. We provide a single illustrative example of applying dynamic SIR, concentrating on the relations of APD to the other vital signs. This variable is typically neglected by the physician. Therefore we are interested in knowing which combinations of the other variables provide important information on the course of APD. We use the data from the same patient as in Section 2.2 (also see [24]). Applying dynamic SIR to the standardized variables yields a three-dimensional reduced regressor space which is mainly spanned by APM and APS, where APD at time t depends strongly on APM at the same time, less but still clearly on APS at time t and APM at times $t-1$, $t-2$, and only slightly on all other variables except for CVP, HR and pulse.

Figure 3 depicts a graphical illustration of these findings. We connect APD to any of the other variables by a line if the simultaneous observation of the latter has a clear impact in the dimension reduction, and draw an arrow towards APD if this is true for a past observation. The strength of the relation (low, medium, high) is interpreted according to the weight the corresponding explanatory variable gets in the dimension reduction step. The exact classification done here is to some extent subjective as we use dynamic SIR as an exploratory tool only. The results are rather similar to those obtained from the partial correlation graph. Both methods identify APM and APS as the most important variables when we want to explain APD.

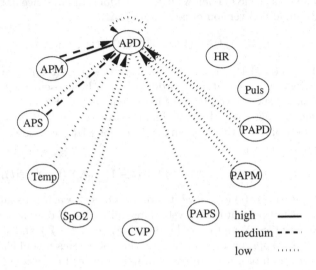

Fig. 3. Relations of APD to other vital signs for one patient derived from dynamic SIR.

Comparing Figures 1 and 3, we see that some low relations of APD to other variables identified by graphical modelling are not found by dynamic SIR and vice versa. This is due to the fact that partial correlation graphs treat both variables symmetrically, whereas dynamic SIR uses an asymmetric regression approach assuming that one of the variables can be explained by the others. For a discussion of the differences between these frameworks in a non-dynamic setting see [25]. For the dynamic setting we just note that dynamic SIR additionally provides some information about the dynamics since it identifies the relevant lags for the relations [24].

Finding a three-dimensional reduced subspace points at nonlinearities since for a linear link function only one direction is needed. This was empirically validated via simulations even in the presence of feedback [19]. In the clinical context, nonlinearities can be due to therapeutic interventions influencing all variables in different ways. A linear factor model as used in Section 3 can therefore only provide a local description of the relations in the steady state. This can be totally satisfactory if detection of relevant changes like intervention effects is the main goal. The possible non-linear nature of the dependence patterns could be analysed further, investigating plots of the variables against the dimension reducing directions [26].

5   Conclusion

Partial correlation graphs and dynamic SIR are useful tools to detect patterns of dependencies in multivariate time series by statistical data analysis. The insights gained by these methods can be useful to improve online monitoring of vital signs since they allow an improved application of methods for dimension reduction. Either a suitable subset of important variables can be selected based on the data analysis, or the information obtained can be used to enhance methods such as principal component analysis [11] or factor analysis for time series if we are able to identify e.g. subsets of closely related variables. In our experience the results of the statistical data analysis agree with medical knowledge. Therefore, we expect to gain new insights when applying these methodologies to other variables, for which we have less medical knowledge so far.

In particular, we have found evidence that specific partial correlation patterns represent specific clinical states in the critically ill as the findings presented here are in good agreement with medical knowledge on the causes and symptoms of distinct states. This can be useful for reaching deeper insights into the causes of clinical complications. In view of the high sampling rates of modern equipment partial correlation analysis could even be useful to detect these complications online in the time critical situations on the intensive care unit.

Acknowledgements. We appreciate the suggestions by a referee as they made the paper more readable. The financial support of the Deutsche Forschungsgemeinschaft (SFB 475, "Reduction of complexity in multivariate data structures") is gratefully acknowledged.

References

1. Imhoff, M., Bauer, M.: Time Series Analysis in Critical Care Monitoring. New Horizons **4** (1996) 519–531
2. Brillinger, D.R.: Remarks Concerning Graphical Models for Time Series and Point Processes. Revista de Econometria **16** (1996) 1–23
3. Dahlhaus, R.: Graphical Interaction Models for Multivariate Time Series. Metrika **51** (2000) 157–172
4. Gather, U. Imhoff, M., Fried, R.: Graphical Models for Multivariate Time Series from Intensive Care Monitoring. Statistics in Medicine, to appear
5. Li, K.-C.: Sliced Inverse Regression for Dimension Reduction (with discussion). J. Amer. Statist. Assoc. **86** (1991) 316–342
6. Whittaker, J.: Graphical Models in Applied Multivariate Statistics. Wiley, Chichester (1990)
7. Cox D.R., Wermuth N.: Multivariate Dependencies. Chapman & Hall, London (1996)

8. Lauritzen, S.L.: Graphical Models. Clarendon Press, Oxford (1996)
9. Edwards, D.: Introduction to Graphical Modelling. Second Edition. Springer, New York (2000)
10. Dahlhaus, R., Eichler, M.: Causality and Graphical Models in Time Series Analysis. Preprint, Department of Mathematics, University of Heidelberg, Germany (2001)
11. Brillinger, D.R.: Time Series. Data Analysis and Theory. Holden Day, San Francisco (1981)
12. Dahlhaus, R., Eichler M.: Spectrum. The program is available at http://www.statlab.uni-heidelberg.de/projects/graphical.models/
13. Granger, C.W.J., Hatanaka M.: Spectral Analysis of Economic Time Series. Princeton Press, Princeton (1964)
14. Fried, R., Didelez, V.: Decomposition and Selection of Graphical Models for Multivariate Time Series. Technical Report 17/2002, SFB 475, University of Dortmund, Germany
15. Peña, D., Box, G.E.P.: Identifying a Simplifying Structure in Time Series. J. Americ. Stat. Assoc. **82** (1987) 836–843
16. Gather, U., Fried, R., Lanius, V., Imhoff, M.: Online Monitoring of High Dimensional Physiological Time Series - a Case-Study. Estadistica, to appear
17. Ombao, H.C., Raz, J.A., von Sachs, R., Malow, B.A.: Automatic Statistical Analysis of Bivariate Nonstationary Time Series - In Memory of Jonathan A. Raz. J. Amer. Statist. Assoc. **96** (2001) 543–560
18. Kano, M., Hasebe, S., Hashimoto, I., Ohno, H.: A New Multivariate Statistical Process Monitoring Method Using Principal Component Analysis. Comput. Chem. Eng. **25** (2001) 1103-1113
19. Becker, C., Fried, R., Gather, U.: Applying Sliced Inverse Regression to Dynamical Data. In: Kunert, J., Trenkler, G. (eds.): Mathematical Statistics with Applications in Biometry. Festschrift in Honour of Siegfried Schach, Eul-Verlag, Köln (2001) 201–214
20. Cook, R.D.: Graphics for Regressions With a Binary Response. J. Amer. Statist. Assoc. (1996) 983–992
21. Imhoff, M., Bauer, M., Gather, U., Fried, R.: Pattern Detection in Physiologic Time Series Using Autoregressive Models: Influence of the Model Order. Biometrical Journal, to appear
22. Karlsen, H.A., Myklebust, T., Tjostheim, D.: Nonparametric Estimation in a Nonlinear Cointegration Type Model. Discussion Paper, SFB 373, Berlin, Germany (2000)
23. Tong, H.: Non-linear Time Series. A Dynamical System Approach. Clarendon Press, Oxford (1990)
24. Becker, C., Fried, R.: Sliced Inverse Regression for High-dimensional Time Series. In: Opitz, O., et al. (eds.): Proceedings of the 25th Annual Conference of the German Society for Classification. Springer-Verlag, Berlin Heidelberg New York, to appear
25. Wermuth, N., Lauritzen, S.L.: On Substantive Research Hypotheses, Conditional Independence Graphs and Graphical Chain Models. J. R. Statist. Soc. B **52** (1990) 21–50
26. Chen, C.H., Li, K.C.: Can SIR be as Popular as Multiple Linear Regression? Stat. Sinica **8** (1998) 289–316

Author Index

Lecture Notes in Artificial Intelligence (LNAI)

Lecture Notes in Computer Science